EDMOND BROWN

Newton, Maxwell, Einstein

What Were They Thinking?

outskirtspress
DENVER COLORADO

Outskirts Press, Inc.
http://www.outskirtspress.com

ISBN: 978-1-4787-5981-2

Outskirts Press and the "OP" logo are trademarks belonging to Outskirts Press, Inc.

Table of Contents

Part 2-Maxwell

Part 1
Newton

Chapter 1
Gravity

In the year 2000 *Time* magazine chose Albert Einstein as the man of the century. It was a century with two world wars, great generals, great statesmen, and outstanding humanitarians. If *Time* had existed in the seventeenth century, there is little doubt that the editors would have chosen Sir Isaac Newton for similar praise. What was so significant about their theories? What bearing did they have on our lives? It seems as though only certain specialists can appreciate their contributions. Newton and Einstein were physicists. Is it possible for someone outside the field of physics to understand and appreciate their efforts?

These men were geniuses, and it is reasonable to assume that their theories are so complex that special training is required to understand them. Actually, their genius does not lie in the complexity of their thinking. It has more to do with their intellectual bravery. They were willing to give up certain ideas that the rest of the world took for granted. In the case of Einstein, it was the notion of time. If there were clocks all over the universe, it is generally thought that it would be possible to decide whether two events took place simultaneously. It turns out that this apparently obvious statement is not correct. A patient reader will see that the research of James Clerk Maxwell, in the nineteenth century, makes the commonsense notion about simultaneity suspect.

Before Newton's publication of his work, it was generally believed that the laws governing astronomical bodies were different from those dealing with earthly phenomena. Heavenly bodies, as they were called, were in perpetual motion. There must have been angels carrying the planets on their paths around the sun. Newton's intellectual bravery had to do with his willingness to dare think that there was a connection between falling bodies on the earth and the motion of the planets. He dared change the meaning of the words *velocity* and *acceleration* from their usual definitions to ones that would accommodate his theory. The simplicity of the equations that encapsulate his theories is amazing. These theories opened the doors to hundreds of physicists and astronomers, who applied them to all sorts of phenomena. The theory can be applied to the motion of gyroscopes, the propagation of sound waves, bridge building, and the weather, to name a few.

It is possible to get a good understanding of what these great men did by trying to imagine their thoughts as they tried to unravel some mystery. Their reasoning can be understood once prejudices are dropped. There are certain words that have specialized meanings to scientists, words like *velocity* and *acceleration,* for example. It is necessary to understand the reason for the change in order to understand the theory. Newton had a good reason for changing the meaning of these words. He was preoccupied with the motion of the planets. His laws of motion and of gravitation very likely resulted from his attempt to explain their orbits. From this it can be seen that physical laws are the outcome of an adventurous mind trying to solve a difficult puzzle. The connections between the research of Newton, Maxwell, and Einstein have led to an enormous cascade of physical thought. They also make an interesting story.

Newton's Laws

Isaac Newton was born in a small town in England, on Christmas Day in 1642. By that time, Magellan (1480–1521) had circumnavigated the earth, and Copernicus (1473–1543) had published his work indicating that the planets traveled in circular orbits about the sun. Subsequently, careful observations by the Danish astronomer Tycho Brahe (1546–1601), which were analyzed by his lab assistant, Johannes Kepler (1570–1630), indicated that planetary orbits were elliptical. Educated men understood that the apparent motion of the sun during the course of the day was a consequence of the earth's spin.

The scientist who very likely had the most influence on Newton was Galileo Galilei (1564–1642). Galileo, who died in the year of Newton's birth, was famous as a scholar and experimentalist. He is often credited with inventing the telescope, a doubtful claim, but he did develop it independently and used it to observe the moons of Jupiter. He publicized the idea that the earth was not the center of the universe, as was being claimed by the Catholic Church. It was just another planet, orbiting the sun, like the others. In fact, this idea led to his becoming a victim of the Inquisition. His later years were spent under house arrest by the church. The claims of Galileo were not essentially different from those of Copernicus, Tycho, and Kepler. However, the power of his influence was sufficient to convince many scholars of the validity of these claims, in spite of the attempts by the church to suppress them.

To Newton, Galileo's astronomical studies were less important than his work with gravitation. The ancient Greek philosopher Aristotle (384 BC–322 BC) had made the pronouncement that gravity was responsible for heavy bodies falling faster than light ones. Aristotle was such an authority on so many topics that almost two thousand years elapsed before anyone questioned the statement. What Galileo demonstrated was that gravity was not responsible for the differences in the motion of heavy and light bodies. If it weren't for the fact that air resistance has a bigger effect on a feather than a coin, they both would fall in precisely the same manner.

Try This Experiment

Galileo's conclusion concerning falling bodies can be made quite plausible. It isn't necessary to make use of a vacuum chamber or a pump. A bit of paper would fall as fast as a heavy book, if it weren't for the fact that air resistance has a bigger effect on the paper than it does on the book. To show this, try the following experiment: Take a book and hold it flat side up. Place a bit of paper on top of the book so that the book will run interference for the paper when they fall. Now drop the book. Hopefully, you've placed it over a bed or pillow for a safe landing. The paper will stay with the book. You will not notice any difference in the motion of the two. Why hadn't anyone done that in two thousand years? Perhaps most people don't question authority.

Galileo is one of the greatest scientists in history. He set a precedent for questioning authority and relying on carefully conceived experiments. His work set the stage for Newton. It is not known exactly whose influences shaped Newton's thinking the most, but he must have pondered about gravity and planetary orbits for a long time before he published his theories about motion.

Newton was also aware of the work of Rene Descartes (1596–1650), a mathematician, philosopher, and scientist who had stated what later came to be known as Newton's first law. The statement is that a body's natural state of motion is in a straight line at constant speed. Descartes had also introduced algebraic methods into geometry and introduced what are today known as Cartesian coordinates. These contributions will be shown to have played a role in Newton's deliberations.

Newton might have been thinking about planetary motion when he came up with his three laws of motion. Let's try to imagine what his thoughts were in light of what he eventually published. The story is told that he was influenced by the fall of an apple to think about gravity. Is it possible that a gravitational force toward the sun could cause a planet to go around it in a circle, or an ellipse? What did Newton already know about gravitational pulls? Galileo's work made it appear that all objects would fall in precisely the same way in a vacuum, regardless of size or weight. Kepler had found that the time it took for a planet to complete an orbit about the sun depended only on the planet's distance from the sun. Planets come in all sizes, and there didn't seem to be any dependence of planetary motion on size or mass. This behavior is consistent with Galileo's observations about gravity. No other mechanism suggested itself.

Galileo's work indicated that when an object falls, its speed increases at a regular rate. This is equivalent to motion with constant acceleration. Let's get in Newton's head once more. Is the acceleration constant because the force on it is constant? It is likely that the acceleration would be greater if the force were greater. Is there also a connection between the way things fall on earth and the way the planets move about the sun? If you throw a rock, it travels in a curved path. It hits the ground eventually. The harder you throw, the farther it goes before it hits the ground. Is it possible, if you threw it hard enough, that the curvature of its path would be less than the curvature of the spherical earth? If so, it would be in orbit about the earth! It was an exciting idea! It was not obvious how to deal quantitatively with this possibility.

Path of thrown rock

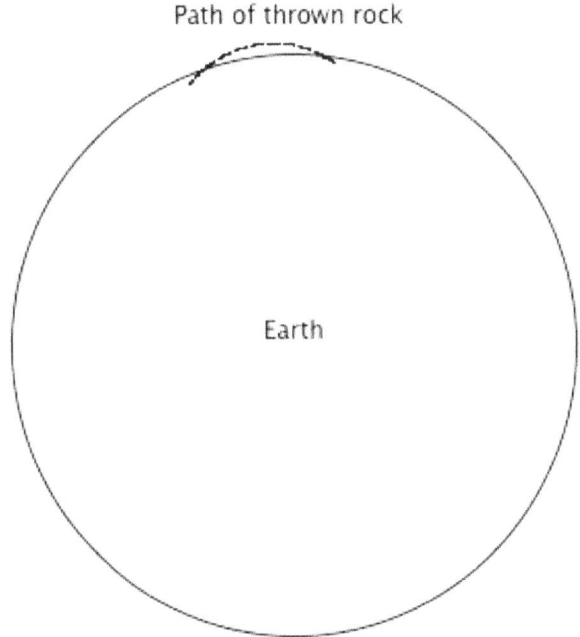

Earth

How to deal with the curved trajectory of a rock in flight? Galileo had studied the vertical motion of an object when dropped. Is it possible that the downward motion of a rock traveling in an arc is the same as it is when dropped? What would the rock's motion look like to someone who tried to run directly below it while it was in flight? Would that person be running at constant speed in a straight line? From our vantage point today, that seems reasonable. If you were on an ocean liner traveling at 30 knots and decided to throw a rock straight up, it would return to your hands (if it were sheltered from the wind). The motion doesn't seem to be influenced at all by the speed of the boat. It would be straight up and down from the view of passengers on the boat but not from those on the shore. The possibility of separating out motion in two dimensions, vertical and horizontal, and treating each one separately, must have been appealing to a mathematician of Newton's caliber.

The idea is illustrated in the photograph below of a plane dropping bombs while in steady horizontal flight. Before the wind has a chance to affect the motion of the bombs, they remain vertically below the plane. From the pilot's viewpoint, the bombs have no horizontal motion at all.

Coordinate Systems and Reference Frames

Newton knew the earth was spinning about an axis. Planetary motion is very complicated as viewed from the earth. Astronomers prefer to describe motion with respect to the stars. If you look up on a clear night, you will see constellations that seem to be fixed in place, as though they are embedded in a celestial dome. From our point of view, here on Earth, the dome is rotating. Planetary motion is easier to understand if the dome is considered as being still, while the earth is spinning beneath it. From the vantage point of this dome, planetary orbits look flat, approximating a circle, with the sun at the center. According to Kepler, it would be more precise to call the orbits elliptical, but it is simpler to deal with circles first.

Specifying planetary motion with respect to the earth presents other problems. The usual words that are used to describe direction, such as *northward*, *up*, or *down* can be misleading. Two people standing near, but on opposite sides of, the North Pole will be looking in opposite directions when both are facing due north. The same is true for two people facing east. Down to an Australian is not even nearly in the same direction as down to a Canadian. The words *horizontal* and *vertical* may have meaning locally but are misleading on a global scale. A person looking at the North Star from a point on the equator is looking horizontally. Another person viewing the same star from the North Pole is looking vertically. They are looking, almost exactly, in the same direction.

Because of the fact that the words *horizontal* and *vertical* don't correspond to fixed directions in space, the description of the motion of the bombs, in the photograph above, is not precise. If the plane were to fly around the earth, it would be clear it was not flying in a fixed direction. Moreover, if you were seated on the celestial dome, you would see that the plane was not traveling in a straight line, even if it were not moving with respect to the earth at all.

If an exact formula is wanted for describing motion, it would be simplest to express the motion with respect to a coordinate system embedded in the celestial dome. There are many possible alternatives for the choice of coordinates. On the earth, the position of an object is usually specified in terms of latitude, longitude, and altitude. The motion of an object can then be expressed by specifying these position coordinates in terms of the time. Descartes invented a more useful coordinate system. These Cartesian position coordinates are shown in the figure below. In actual usage, it would be necessary to provide units, such as feet, meters, and miles. The advantage of such a system, for Newton, was that the simplest motion of all, the motion of a body subject to no forces, could be expressed very simply. By a suitable orientation of the coordinate system, such motion could be expressed in such a way that only one coordinate depended on the time. The others would be constant.

Cartesian Coordinate System
The origin is at the intersection of the axes (coordinates 0, 0).

The use of negative numbers in Cartesian coordinates allows us to extend the coordinates as far as necessary. It is similar to the use of negative numbers to specify temperatures colder than zero degrees. For the general case of motion in three dimensions, a third coordinate perpendicular to the plane of the paper is necessary. A point in three-dimensional space could then be specified by the coordinates (x, y, z). Newton could imagine such a coordinate system embedded in some rigid framework. Such a structure could be used to specify the position of any point in the universe. The system is called a frame of reference. A body may be stationary as seen from one reference frame but be moving as seen from another. For example, if a coordinate system were embedded in a subway car, people seated in the car would be considered stationary while the subway stations could be moving.

Today, scientists are leery of extending coordinate systems outwardly without limit. Such an extension makes assumptions about the nature of space. It assumes that space is flat and infinite, in some sense. Scientists don't want to make the kind of mistake that people made when they assumed the earth was flat. To make this point a little clearer, try to imagine a straight-line journey through space in which you eventually come back to your starting point. If you find that difficult to do, you might be suffering from the same lack of imagination that plagued the believers in a flat earth. However, you cannot blame Newton for making the assumption that space is flat, at least in our neighborhood in the solar system.

Readers should be warned that mathematical equations are used throughout this book. They are included despite the warning by Stephen Hawking that an author loses half the potential readership for every equation included in a book. Physics and mathematics are closely intertwined, and to tell the story without math is to cheat a little. However, let the reader be reassured that it is possible to skim over the equations, and even ignore some of them, without much harm. In no case is the reader required to do any calculating, although a few exercises are suggested. Mathematical proofs are included for those readers who like to see how everything fits together. You should feel free to ignore these.

The Problem with Defining a Rate

(A Prelude to the Calculus)

Newton's thoughts about motion were completely new. Galileo's gravitational studies of motion had to do with motion in a straight line. They suggested that the speed of a freely falling object increased at a constant rate. How did Newton deal with the rate of change of speed? The notion of speed itself presents a mathematical problem. It is a rate, and rates involve a ratio, or a quotient, like miles divided by hours. If you are in a car going 60 mph (miles per hour), you expect to cover a distance of one mile in one minute. It is necessary to measure both a distance and a time to check whether your speedometer is accurate. How do you know if the speed has varied during that time? You would have to check by taking shorter time intervals and measuring shorter distances and computing the ratio again. But you still don't know for sure how fast you are going at any instant. An instant has no duration. You travel zero distance in zero time; division of zero by zero is a no-no to a mathematician. It is not determinate.

Newton wanted to deal with the rate of change of a rate. He asked himself, "Suppose I had a record of the position of a particle (an object small enough to be approximated by a single point) at all times. Is it possible to assign a speed to it at every instant along its path?" Newton was willing to consider motion in a straight line, since he believed he could focus on one Cartesian coordinate at a time. "If I could get a record of the speed at all times by some procedure, then it should be possible to use a similar procedure to get the rate of change of speed." Perhaps Newton was thinking of Galileo's conclusion that the distance covered by an object dropped from rest is proportional to the square of the time. This can be put in the form of an equation, as

$$x = At^2,$$

where x represents the distance fallen in time t. The symbols just mean that this distance is the product of some constant with the square of the time of fall. The square of the time, in turn, is just the product of t with itself. For example, when t takes on the values 1, 2, 3, 4, 5 for the time of fall in seconds, say, then t^2 takes on the values 1, 4, 9, 16, 25. The choice of the constant A depends on the units chosen for time and distance. If t is measured in seconds and x is measured in feet, A is approximately 16. It will be different if x is measured in meters. You can make it whatever number you want by inventing your own system of units. For example, A could have the value 1.00 by choosing the unit of distance to be the distance that the object falls in the first second. The unit of distance is then approximately 16 feet. Newton might have tried to get an estimate of the speed at a given instant, as well as its rate of change, by tabulating this information. He would have used the equation,

$$x = t^2,$$

in which t is measured in seconds, and distance is measured in units approximately 16 feet long. Suppose he tabulated this information and made use of an approximate procedure to get the speed and the acceleration as shown below.

t	0	1	2	3	4	5	6	7	8
x	0	1	4	9	16	25	36	49	64
v	1	3	5	7	9	11	13	15	
a	2	2	2	2	2	2	2		

Explanation: The top row indicates the value of the time of fall in seconds after being dropped. The next row is the distance fallen, according to the equation $x = t^2$ in the assumed unit of distance. Thus, the distance fallen at t = 5 seconds is 25 units, or approximately 25 times 16 = 400 feet. The third row, labeled v, is a crude estimate of the velocity at the given time. The estimate for v at t = 3, for example, is 7 units of distance per second. It is probably a better estimate for the speed at t = 3.5 because it is obtained by taking the distance covered between t = 3 and t = 4, namely 16 - 9 = 7, divided by the elapsed time of one second. Do not be concerned at this point that the value of v is not completely meaningful. The best that can be said for v is that it is a good estimate for the speed at a time between t = 3 and t = 4. The last row is labeled a for acceleration. Again it is an estimate. The value at t = 3 is obtained by finding the change in speed between t =3 and t = 4, then dividing by the elapsed time. The fact that it is constant suggests that it might be an accurate value in spite of the fact that the values of v are approximate. Thus, the table suggests that the acceleration is constant at a value of 2 units per second per second. In other words, the speed goes up by 2 units per second each second. Since 2 units is approximately 32 feet, this can be written as a = $32ft/sec^2$. This is an abbreviated notation. Although it is read as 32 feet per second squared, it just means that the speed increases by 32 ft/sec each second.

Newton recognized the inadequacy of the tabular method for obtaining the speed at a given moment. All that has been obtained is the average speed in a given time interval, and then that average is used as the speed at the beginning of the interval. Clearly the error would have been smaller if shorter time intervals had been used. For example, if the speed at t =3 had been estimated by using the average speed in the interval between 3 and 3.5 instead of the interval between 3 and 4, it would have been better. The table below is constructed by the same method as before, except that the time is given at every half second.

t	0	.5	1.0	1.5	2.0	2.5	3.0	3.5	4.0	4.5
x	0	.25	1.0	2.25	4.0	6.25	9.0	12.25	16.0	20.25
v	0.5	1.5	2.5	3.5	4.5	5.5	6.5	7.5	8.5	
a	2.0	2.0	2.0	2.0	2.0	2.0	2.0	2.0		

The value for v at t = 3 is obtained by noting that in the interval between 3 and 3.5, it travels 12.25 - 9 = 3.25 units in one half second.

This yields an average of 6.5 units per second instead of 7. No matter how short an interval that is chosen, the tabular method is not reliable. Newton found a better method than the tabular one.

Differential Calculus

Consider the same motion, namely
$$x = t^2,$$
and focus your attention on one particular time, t. This is equivalent to consideration of one column of the tabular method. Suppose the symbol, s, denotes the number of seconds between columns in the table. In other words, an algebraic symbol replaces the numbers 1 and 0.5 that were used previously. The column in the table next to t is t + s. The entry for x in this column of the table is (t + s)². A formula is needed for this, which turns out to be
$$(t+s)^2 = t^2 + 2st + s^2.$$
The proof of this formula is simple, as will be shown shortly. The procedure Newton followed is the same as that of the tabular method. The distance traveled in the time between the time t and the time t + s is
$$(t^2 + 2st + s^2) - t^2 = 2st + s^2 = s(2t + s).$$
In order to get the average speed in this interval, it is necessary to divide by s. The result is
$$v = 2t + s.$$
This result can be checked for the two tables considered earlier. In the first table, the formula is v = 2t + 1. In the second table, it is 2t + 0.5. You can check back to verify that this is indeed what was found earlier. For t = 3, the formula yields 7 and 6.5. The difference between the algebraic and the tabular methods is that the algebraic method specifies the result for any conceivable table. If an interval of 1 millionth of a second had been taken the formula would have been, v = 2t + 0.000001. It should be obvious that the formula v = 2t is an accurate measure of the velocity of the body at time t. The letter v will no longer denote an average value. If

you have followed the arguments given here, you know what it means to differentiate with respect to time (alternatively: to take a time derivative.) The idea used by Newton, of taking a limit, avoided the problem of dividing zero by zero. It turned out to be an extremely useful idea in much of mathematics.[*]

---

[*] Newton shouldn't be given sole credit for the invention of calculus. The mathematician Gottfried Leibniz (1646–1716) invented this technique independently. He made valuable contributions to the subject. It is his notation that is in most common use today. It is the one used in this book.

In order to prove the formula used earlier for the calculation of the square of a sum of two quantities, consider the diagram below. Each algebraic symbol denotes a length rather than a time. They are just numbers, as long as the units have been assumed (inches or feet, perhaps). The large square has a side of length $(t + s)$. Its area is thus $(t + s)(t + s)$, which can be written as $(t + s)^2$. This area is made up of two smaller squares and two rectangles. The areas of the enclosed squares are t^2 and s^2. The area of each rectangle is st. The area of the whole figure is equal to the sum of the areas of its parts, so that

$$(t + s)^2 = t^2 + s^2 + 2st.$$

Question: Can you draw a figure to show that $t^2 + 2st = t(t + 2s)$?

t   S

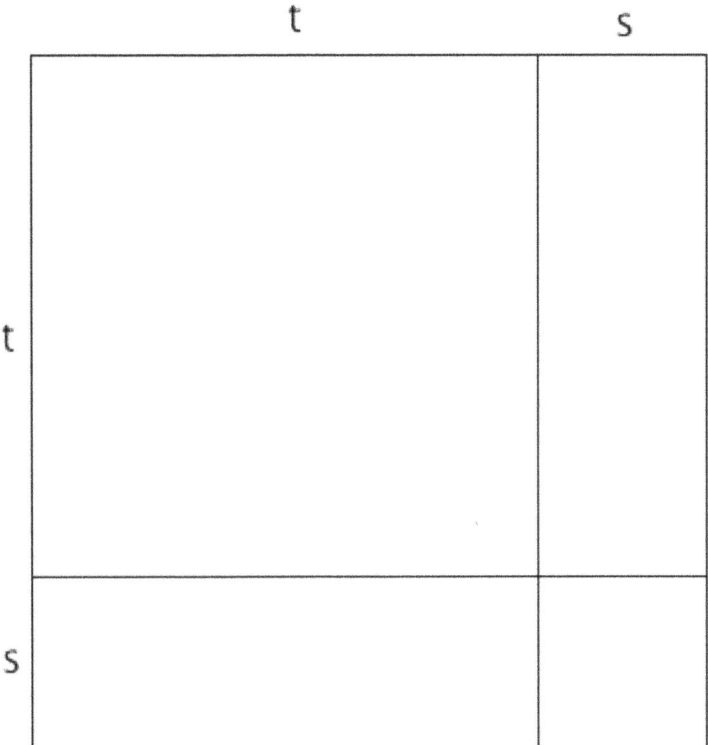

t

S

Newton found a method for calculating the velocity of an object at any instant of time. It assumed that a record of its position at all times was available. The same method could be used to calculate the body's acceleration from its speed. It involved a derivative with respect to time. In our present-day notation, the formula is written as

$$a = \frac{dv}{dt}.$$

The letter d designates a differential quantity, similar to the differences used in the tabular method. Thus, dv designates the change in velocity in the short time interval dt. These differential quantities are assumed to approach 0. The formula yields the same result that was found by the tabular method, namely a = 2. This is equivalent to an acceleration of 32 ft/sec^2.

Newton must have realized that Galileo's experiments with falling bodies was limited to motion in a straight line and restricted to a region near the earth's surface. It wasn't directly connected to the kind of motion he wanted to consider, but it must have made him aware that it was the acceleration that was the quantity of interest. Perhaps the acceleration of a falling body was constant because the force on the falling body was constant. After all, the weight of a body didn't change from point to point, at least not noticeably in Newton's day. The problem, however, was that the acceleration didn't seem to work for planetary motion. The planets were traveling at nearly constant speed. If acceleration had to do with the rate of change of speed, the acceleration of the planets would be nearly zero.

A graphical plot of v versus t, for the motion just considered, is a straight line, as shown below. The slope of the line is a measure of the rate at which v is changing and is, thus, proportional to the acceleration.

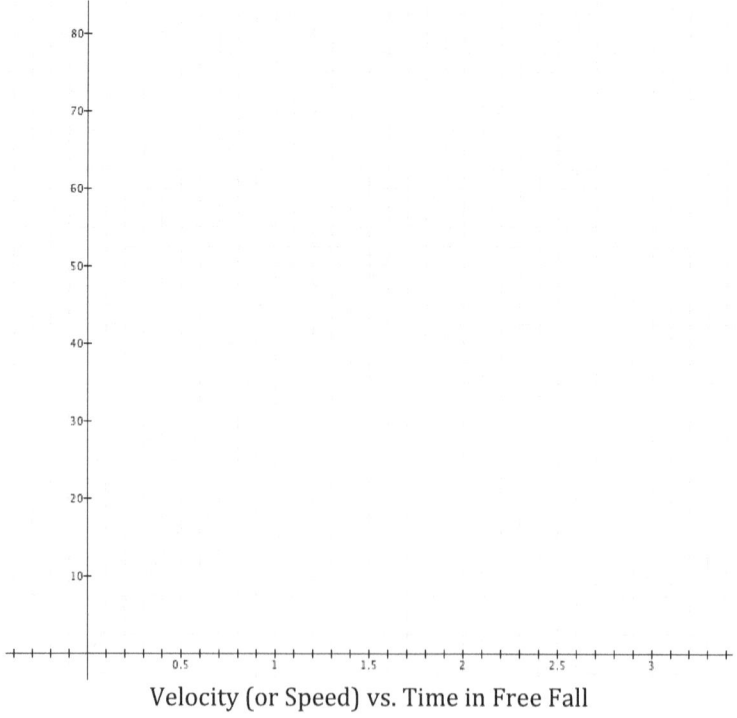

Velocity (or Speed) vs. Time in Free Fall

The Difference between Velocity and Speed

What would a plot of speed versus time look like if the object was thrown upward instead of falling from rest? Clearly its speed would be going down on the way up. It would be decelerating. A graph of speed versus time might look like the one sketched below. The speed is dropping as the particle approaches the top of its path. After that, it is expected to rise. There is a sudden change in the slope of the graph when v reaches 0. This corresponds to the top of the path, at the time when the body is turning around. A layman might say that the acceleration had changed abruptly. To a mathematician of the caliber of Newton, this abrupt change in slope is misleading.

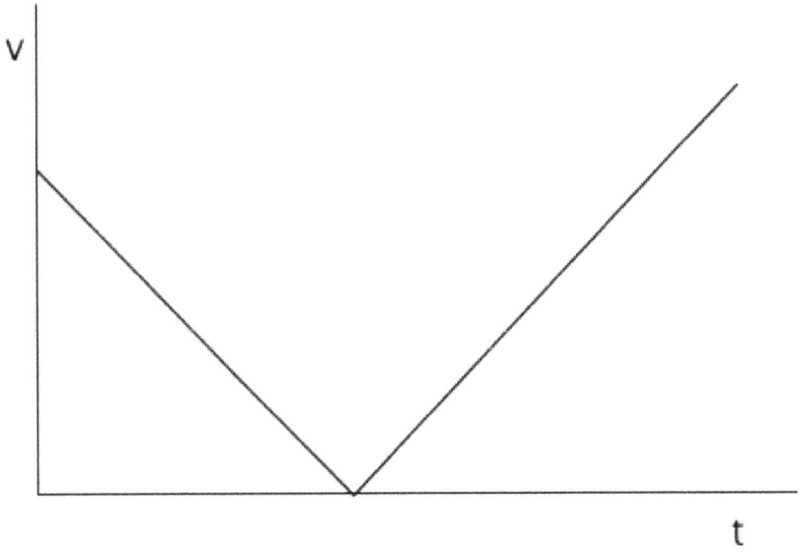

Graph of Speed (not Velocity) vs. Time

The reason for the sudden change comes from the fact that acceleration is defined as the rate of change of speed. Speed, on the other hand, is not really a rate of change of anything. It is never negative. The speedometer in a car starts at 0. It specifies the magnitude (the size) of a rate of change but not the direction. A rate of change can be negative as well as positive. The difference between the two can be important. Consider the rate of change of your bank account, for example. If you are told it is ten thousand dollars a year, wouldn't you want to know whether this rate is positive or negative? If the altimeter reading on your plane is changing very rapidly, you certainly would want to know if the rate is positive.

The words *velocity* and *speed* have been treated as synonymous. The physicist makes a distinction between these two words. If the upward direction is treated as being positive, for example, motion in the downward direction can be specified by a negative velocity. Thus, if the velocity had been plotted, rather than the speed, in the previous graph, there would be no break in the slope of the line. It would be one continuous straight line. The right-hand part of the graph would be replaced by its mirror image, extending below the x-axis. The acceleration for this motion is negative, or downward, throughout the motion. Such a motion has constant acceleration. There is no need to use the word *deceleration* to describe a motion, since a given acceleration can be associated with speeding up or slowing down. Since Newton's time, physicists have defined acceleration as the rate of change of velocity, not the rate of change of speed.

The change in the meaning of the word *acceleration* must have opened up a world of possibilities for Newton. It was now possible for the planets to be accelerating even though their speed was nearly constant. He needed a definition of acceleration that was valid for all sorts of motion. He also needed to examine the implications of his new way of thinking. If acceleration was the rate of change of velocity, not speed, he needed a definition of velocity that was valid for motion in three dimensions.

It is necessary to reconsider the earlier motion, in which x denoted the distance an object fell. Like the word *speed*, the word *distance* is restrictive. It cannot take on negative values. It is more appropriate to think of x as a measure of position. In the one-dimensional motion of a vertical fall, it could represent the altitude of a body, measured from some reference point. If g were used to designate the acceleration due to gravity, the equation of motion would be as follows:

$$x = x_0 + v_0 t - \frac{1}{2} g t^2.$$

Here, x_0 is the starting position coordinate; v_0 is the initial velocity (positive if upward). The last term is negative because g is conventionally listed as a positive number, whereas the acceleration is negative (downward). In this case, the velocity is specified by the formula

$$v = \frac{dx}{dt} = v_0 - gt.$$

Newton realized these equations were approximate. The earth is spinning, and g is not really a constant. He needed to use a frame of reference in which the celestial dome was fixed in which to embed his coordinate system. He relied on the work of Rene Descartes and chose three Cartesian coordinates to specify the position of a planet. It was reasonable to ignore the motion of the sun and to choose the origin of his coordinate system at the solar position. Three numbers are needed to specify the velocity once the time dependence of the position coordinates are known. They are given by

$$v_x = \frac{dx}{dt}, \quad v_y = \frac{dy}{dt}, \quad v_z = \frac{dz}{dt}.$$

In a similar way the acceleration has three components given by

$$a_x = \frac{dv_x}{dt}, \quad a_y = \frac{dv_y}{dt}, \quad a_z = \frac{dv_z}{dt}.$$

Once Newton had changed the meaning of the words *velocity* and *acceleration*, it was necessary to find rules that described the motion of bodies. He was guided by a statement made earlier by Descartes.

Newton's First Law of Motion

Newton thought of gravity as a pull, a force. For the apple, it was the earth that was doing the pulling; for the planets, it was the sun. What is the relation between force and motion? A falling object had almost constant acceleration in Galileo's experiments because the force on it was almost constant near the surface of the earth. A massive object like a bull or an elephant needs a much greater force, for a given acceleration, than a bird or a feather. In this context, mass denotes the inertial properties of a body (the tendency to resist a change in velocity). Perhaps the acceleration is inversely proportional to the mass. If motion under gravitational forces doesn't depend on mass, maybe it is because the gravitational force is proportional to the mass. In a sense, the gravitational property of mass and the inertial property of mass cancel, leading to motion that doesn't depend on mass when the only force on a body is gravitational.

What would motion look like if there were no forces on an object at all? Here on Earth, friction is encountered everywhere. Moving objects can be slowed by the underbrush, the air, or a rough terrain. Wouldn't motion look different in a region where there was no friction and far from any gravitational affects? Newton could easily imagine that the planets were traveling through a vacuum, a region devoid of friction. Would a planet travel in a curved path if there were no sun? Which way would it curve? It would very likely move in a straight line and not slow up. The idea wasn't original. Rene Descartes said the same thing years earlier. Nevertheless, it came to be known as Newton's first law. A fairly precise statement might read as follows:

The natural state of motion for a body is to travel in a straight line at constant speed. This is equivalent to the following: The natural state of motion is of constant velocity, or zero acceleration.

Although not explicitly stated, this law implies that you are considering motion as viewed from a specific vantage point, not on the spinning earth. In fact, physicists interpret the first law as implying that there exist reference frames in which bodies subject to no forces will coast in straight lines at constant speed. If one such frame of reference can be found, it is possible to find another. It would be a frame that is coasting at constant velocity with respect to the first and not spinning with respect to it. All these special frames in which the first law is valid are called inertial reference frames. If you were on such a frame of reference, the constellations would not seem to be moving at all. You would have to imagine the sun was very far away so that it didn't affect your coasting through the cosmos at a steady speed in a straight line.

Up to this point, an attempt has been made to get into Newton's head and imagine the kind of problems that were occupying his thoughts. His laws of motion are elegant attempts to answer the questions he was asking himself. However, these laws are deceptive in that every term he used has to be examined carefully in order to be understood.

The Second Law

Newton's second law is one of the most famous laws in all of physics. It is usually written as F = ma and stated as "force equals mass times acceleration." However, this formulation is misleading. It is better to state it in the form

$$a = F/m.$$

In words, it becomes "the acceleration of a body equals the net force on it divided by its mass." This is mathematically equivalent to the first way of writing it, but it helps us realize that it is the acceleration of a body that is determined by the net force acting on it, not the other way around. The forces acting on the body are determined by other factors.

A remark about the second law is in order. A constant, k, should be inserted into the equation. Thus, the equation becomes $a = kF/m$. This constant can be eliminated by the proper choice of units. One such system, in common use by scientists, is the MKS system. In this system, the unit of length is the meter (39.37 in), the unit of mass is the kilogram, and the unit of time is the second. In this system, the unit of force is called the newton. A force of one newton is that force that will give a 1 kilogram mass an acceleration of 1 meter/sec^2. A 1 kg mass weighs about 2.2 lbs on the surface of the earth. A 2.2 lb. force will give a 1 kg mass an acceleration of 9.8 meter/sec^2 (32 ft/sec^2). From these facts you should be able to compare the force of 1 pound with that of 1 newton. Hint: The pound is bigger.

It is hard to imagine an equation that looks any simpler than the second law. It turns out to be amazingly subtle and amazingly useful. Let us look first at the case where there are no forces acting on a body. It then follows that the acceleration is zero. It seems as though the first law isn't necessary since the second law tells us the same thing. Remember, however, that the first law deals with special frames of reference called inertial frames. The second law is only valid with respect to these special reference frames. If you use it in any other case, you are either making an approximation or a mistake. Thus, strictly speaking, Newton's Law isn't valid with respect to a reference frame fixed on the earth. It is often a good approximation, however, to use it on the earth for phenomena that last only a matter of seconds. The earth doesn't turn very much in that short a time.

It is important to examine the meaning of each symbol in the second law. They are taken up in detail here.

Force: Of the three terms involved in the second law, force is the most complicated. In Newtonian physics, all forces on a body arise out of interactions with other bodies. There are no exceptions. You probably have heard explanations of why water doesn't spill out of a bucket that is being whirled in a vertical circle in terms of centrifugal force. It is advisable to forget these explanations, as they are almost always wrong. (To be sure, if you want to take the bucket as a frame of reference, it is possible to do so by introducing fictitious forces, such as centrifugal force. For the time being, accept the fact that these forces are fictitious. They have been introduced in order to make use of non-inertial frames of reference.)

Physicists today tell us of four basic forces: strong, weak, electromagnetic, and gravitational. Of these, Newton was only familiar with the last. For our present purposes, it is most useful to divide forces into two categories: 1. contact forces 2. action-at-a distance forces. The only one in the latter category with which Newton concerned himself is that of the gravitational force. The rest of the forces come about because of deformations of bodies in contact with one another. Air resistance can even be considered as such a force. It is caused by large numbers of gas molecules hitting a body each second and deforming as they make contact.

It is important to recognize that force is a vector quantity, which means that it has a direction as well as a magnitude. When a number of bodies are exerting forces on a body, it is necessary to sum these vectors to determine the net force. Forces can be specified in terms of their Cartesian components. The x-component of a sum of forces, for example, is just the sum of the x-components of the individual forces.

Mass: This is a property of a body. It doesn't depend on a body's location. Newton assumed that a bag containing a dozen baseballs would have twice the mass of a bag containing six baseballs. If the accelerations of the two bags are the same, the force on the bigger bag must be twice as great as the force on the other. (Granted, Newton never saw a baseball.) Since the two bags fall in precisely the same way in a vacuum, the weight force on the bigger bag must be double the weight of the smaller one. In fact, the weight of a body must be proportional to the mass, since, according to Galileo, the acceleration of free fall, g, is the same for both bodies. According to the second law, when applied to a freely falling body, then

$$W = mg.$$

This equation is valid whether the body is falling or not. The mass of a body doesn't change unless a body loses or gains material, such as by evaporation or chemical reaction. The weight, which is the magnitude of the earth's attractive force on the body, does change with position, and so does g. Note that the weight force is truly a vector (pointing downward), and so is g. Mass, on the other hand, has no directional properties. It is a scalar.

There are two distinct properties associated with mass. On the one hand, it is a measure of a body's inertia. This means that for a given force, a body with twice the inertial mass of another body will have half the acceleration. On the other hand, it is also a measure of the gravitational force on a body in a given location. A body with twice the gravitational mass of a second body will have twice the weight at a given location. Newton didn't distinguish between the two aspects. Gravitational mass and inertial mass were one and the same. Today, some scientists wonder why such a coincidence should occur. Why doesn't mass matter (no pun intended) when something falls? Experiments have been carried out to tremendous accuracy in recent years. As of this writing, no measurable difference between the two types of mass has been found. No distinction will be made between the two meanings. A scale can be used to compare masses. The kilogram mass has been defined as the mass of a platinum-iridium cylinder stored in Sevres, France. Secondary standards are in use in different countries. There is some concern that the materials aren't stable enough, and some better standards are necessary. Scientists are willing to revise standards when something better becomes available. Such a revision will be encountered when the unit of time, the second, is defined.

Acceleration: Acceleration has been defined as the rate of change of velocity with time. It is assumed that these are measured in some inertial frame. Velocity, in turn, is defined as the rate of change of position with respect to time.

"So, what good are Newton's laws to us, here on the earth, if they don't work because of the earth's rotation?" you could ask. One answer might be that they work well enough for many purposes, because our planet turns so slowly. Another answer might be "If a body's motion with respect to an inertial frame can be measured, it is then possible to specify its motion with respect to the earth." This is the more important answer. It is a straightforward matter to adapt Newton's laws so that they apply to bodies on the earth. Far from treating the earth as a special place at the center of the universe, Newton took us into a new world. You might say that the space age began with him.

Why the Need for Particles in the Second Law

It is necessary to make a slight alteration in what has been written so far. Newton realized that bodies have spatial extent and that all the parts might not be moving in the same way. In a moving automobile, for example, the different parts of the wheels are traveling at different velocities with respect to the ground. The velocity of a body, or its acceleration, has no meaning if different parts of the body are moving differently. In order to deal with such a situation, Newton imagined that it is possible to divide the body into such small parts that variations of velocity within each part are negligible. Newton referred to these bits of matter as particles and applied his laws to each of them. Newton's laws then contain the word *particle* instead of *body*. In Chapter 4, it will be shown that Newton's laws can be applied to an extended body as long as reference is made to the acceleration of a particular point, known as the center of mass of the body. This is useful in showing that the orbits of planets are not affected by planetary rotation. Since a planet is spherically symmetrical, its center of mass is at its geometrical center, regardless of planetary rotation. Planets can be treated as though they were particles, as far as their orbits are concerned.

The Third Law

Newton's third law is simple to state, but it is counterintuitive to many (perhaps most) people. A fairly precise statement might read as follows: "Whenever body A exerts a force on body B, then body B exerts a force on body A that is equal in magnitude but in the opposite direction." Remember, the first word is *whenever*. There is nothing implied about equilibrium. There are no delays. There are no statements to the effect that one body causes the interaction. The two bodies are on an equal footing. There are two and only two. A shorthand statement of the third law is "action equals reaction." However, since the two bodies involved are on an equal footing, it is arbitrary as to which force is the action. The shorthand statement also doesn't emphasize the important fact that the two forces act on (and are caused by) different bodies.

What led Newton to think of the third law? It is interesting to speculate on the type of reasoning he might have used. The motion of the planets, as described by Kepler's laws, is extremely regular, dependent only on the distance from the sun. If it is assumed that planets are subject to internal forces, like those on the earth, associated with storms, avalanches, volcanoes, and earthquakes, they have no discernible effect on the overall motion. As will be seen in Chapter 5, the motion of the center of mass of a body is unaffected by internal forces. This result is a consequence of Newton's third law. If one part of a body could push on another part, without feeling an equal size reaction force, it would be possible to change the motion of the center of mass of the body, without the need for external forces. When a young child attempts to lift herself off the ground by pulling upward on her feet, she is unaware of Newton's third law.

Scalars and Vectors

Physical quantities require numbers and units. A quantity like that of mass requires only one of each, but three numbers have to be specified in the three-dimensional world to specify position, velocity, and acceleration. Actually, these three numbers depend on the frame of reference and the choice of coordinate system. Such quantities are called vectors, while those quantities, like mass, that don't depend on the coordinate system are called sca-lars. Quite frequently the motion of a particle is confined to a plane. In such a case, it is convenient to choose a coordinate sys-tem embedded in that plane. The z coordinate can be ignored, for example, and the vector can be expressed in terms of the x and y components.

Today, physicists often display vectors as arrows. For example, if the components of velocity are given as

$$v_x = -4\,ft/\sec, \quad v_y = -3\,ft/\sec,$$

the velocity vector can be specified by means of an arrow, as in the diagram below. The position vector can be specified by an arrow, from the origin of our coordinate system to the coordinates of the point in question. Boldfaced symbols are used in the text to distin-guish vectors from scalars. Thus, the velocity vector is denoted by the letter **v.** The speed of the particle, v, is just the magnitude of this vector. The Pythagorean theorem, (proven in the next section) can be used to obtain the result

$$v = \sqrt{v_x^2 + v_y^2} = \sqrt{16+9} = \sqrt{25} = 5\,ft/\sec.$$

In the most general case, the formula is

$$v = \sqrt{v_x^2 + v_y^2 + v_z^2}.$$

Note that Newton's second law actually consists of three equations, namely

$$a_x = F_x/m, \quad a_y = F_y/m, \quad a_z = F_z/m,$$

with reference to some Cartesian coordinate system embedded in an inertial frame of reference.

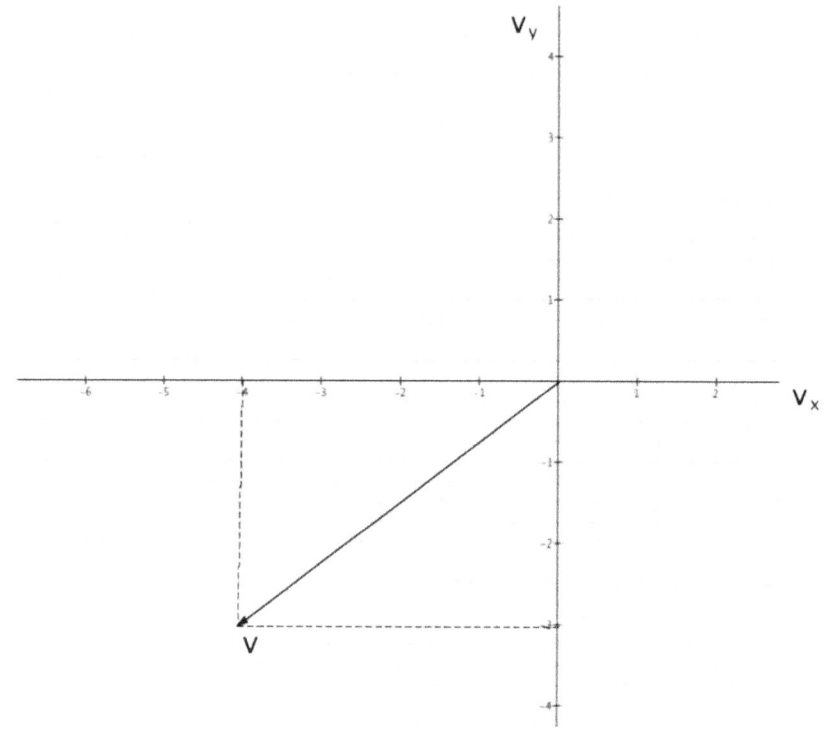

Velocity Vector as an Arrow

Right Triangles

This section is devoted to a few definitions and a few mathe-matical derivations. You might find these helpful in dealing with the more interesting subject matter of the next chapter. If you are not interested in derivations, you might prefer to skim over the equations that define certain trigonometric quantities.

The figure below consists of a rectangle, ABCD, that has been divided into two identical triangles, ABC and ADC. Let's focus our attention on the triangle ADC and note that the Greek letter *theta* (θ) denotes the angle at the vertex A of this triangle. (Physicists often use Greek letters to denote angles.) The side opposite to vertex A, CD, is of length a, and the side adjacent to it, AD, is of length b.

The angle at the vertex D is a right angle. (The angles at all the corners of a rectangle are right angles.) The side opposite this, AC, of length c, is called the hypotenuse of the triangle. Every triangle that contains a right angle is called a right triangle. If this figure were scaled up or down, each length would change by the same factor. However, the angles wouldn't change. Since the ratios of these lengths are unaffected by scaling, the fraction a/c depends only on the angle θ.

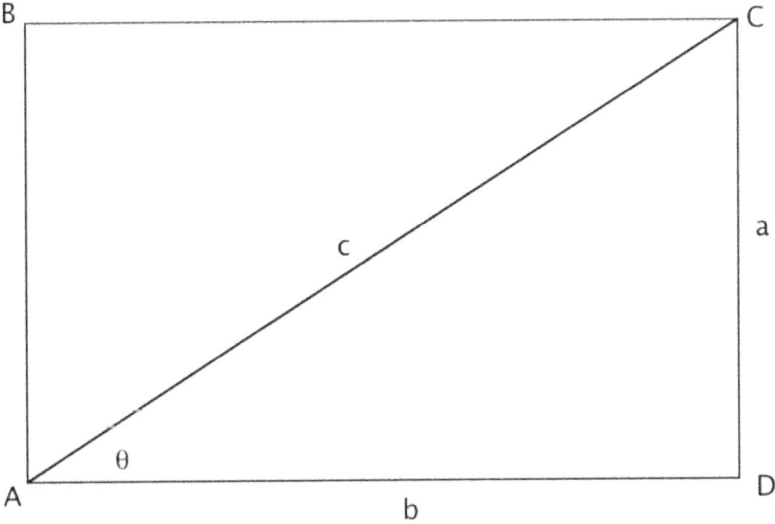

The common unit of angular measure is the degree. It is defined in such a manner that a right angle is 90 degrees, written as 90^0. It follows that the sum of the angles in the rectangle is 360^0, and the sum of the angles in the triangle is 180^0. (Many readers probably already know that the sum of the angles doesn't depend on the presence of right angles.) The fact that the sum of the angles in the two triangles is the same has been used.

There are certain ratios that are used so frequently that they are given names. The ratio a/c, which depends only on θ, is called the sine of the angle θ. The letter e is omitted from the word *sine* when writing this in an equation. Thus, the equation is written as

$$\sin \theta = \frac{a}{c}.$$

The equation for the cosine is written as

$$\cos \theta = \frac{b}{c}.$$

One more ratio is in common use. It is called the tangent and is defined as the opposite over the adjacent, so that

$$\tan \theta = \frac{a}{b}.$$

The definitions given by the previous three equations only work for angles less than 90°. The definitions will be extended in the next chapter so as to be applicable to any angle. The three trigonometric functions are related to one another. Knowledge of one of them determines the other two. In particular, it will be shown that

$$\cos \theta = \sqrt{1 - \sin^2 \theta}, \qquad \tan \theta = \frac{\sin \theta}{\cos \theta}.$$

Note that the superscript 2 is placed where it is in order to denote that the sine of the angle is to be squared, not the angle.

An alert reader might be able to show that the last equation can be derived from the previous equations in this section. The first depends on the Pythagorean theorem.

Exercise: Show from the fact that the sum of the angles in a triangle is 180° the following:

$$1. \sin(90^0 - \theta) = \cos \theta, \text{ and } 2. \cos(90^0 - \theta) = \sin \theta .$$

The Pythagorean Theorem

The mathematical theorem that will be proved here dates back to the interesting Greek mathematician and scholar Pythagoras (570 BC–495 BC). The theorem can be stated very simply in mathematical terms. Referring back to the triangle of the last section, the statement of the law is as follows:

$$a^2 + b^2 = c^2.$$

Numerous proofs have been given, some of which are quite complicated. Two proofs are given here. If one seems too difficult, try the other.

Proof #1.

In the figure below, the original triangle, ADC, is redrawn. A line, DE, is drawn that makes a right angle with the hypotenuse of the original triangle, AC. Let us show the reasoning that allows us to conclude that the angle specified by EDC is the same as the angle CAD, namely θ. Consider triangle ADE. From the fact that the sum of the angles is 180⁰, it is found that the angle ADE + θ = 90⁰. However, angle ADE and angle EDC add up to 90⁰ also, because angle ADC is a right angle. Note that all three triangles, ADE, EDC, and the original one, ADC, contain the same set of angles.

The proof of the Pythagorean theorem now consists of two parts. Note that the area of the original triangle is the sum of the areas of the other two triangles.

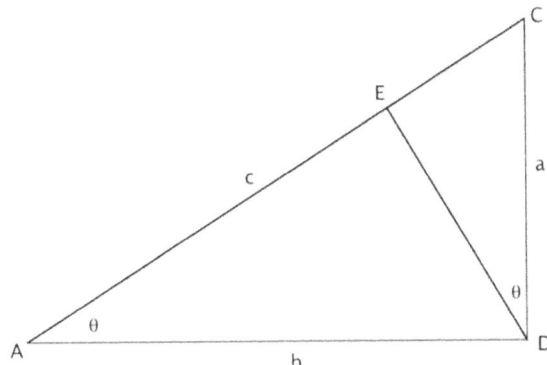

The second part consists of expressing these areas in terms of the hypotenuse and the angle θ. Let's recall that the triangle under consideration was obtained by dividing a rectangle into two equal parts. Since the area of a rectangle is obtained by multiplication of the lengths of two adjacent sides, it follows that

Area of ADC = ab/2.

This can be expressed in terms of the hypotenuse by use of the relations

$$a = c \sin \theta, \qquad b = c \cos \theta,$$

so that

Area of Large Triangle = c²sinθcosθ/2.

Since there is nothing special about the large right triangle, a similar formula holds for any right triangle in which one of the angles is θ. All such triangles are similar, differing from one another only in scale. This result is applicable to the three right triangles in the figure, since they all contain the same three angles. In other words, the area of each triangle is given by a formula of the form area = k times the square of the hypotenuse, where $k = \sin \theta \cos \theta / 2$. The rest of the proof is simple:

Area of ADC = Area of DCE + Area of ADE

$$kc^2 = ka^2 + kb^2.$$

Division by k leads to the Pythagorean theorem;

$$c^2 = a^2 + b^2.$$

Proof #2.

Consider the original triangle, ADC, and imagine that you have cut out four pieces of cardboard exactly like it. Assume you have assembled these four triangles as in the figure below.

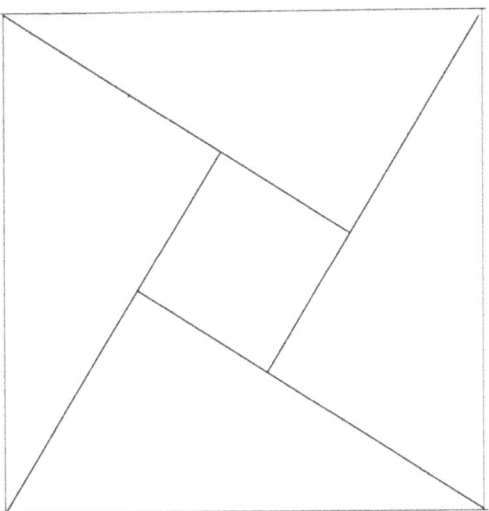

Readers should be able to convince themselves that the four triangles assemble into a square with a square hole. The larger square
has an area given by c². The square hole has a side of length (b - a),
under the assumption that b is larger than a. The hole has an area
given by (b - a)². Recall the formula for (t + s)² and replace t with b
and s with -a. Thus:

$$\text{Area of hole} = b^2 + a^2 - 2ab.$$

The area of the four triangles, which make up the rest of the large
square, is just 2ab. Once again it is found that

$$c^2 = a^2 + b^2.$$

Consider the given triangle, ADC, once more and note that

$$\sin\theta = \frac{a}{c}, \qquad\qquad \cos\theta = \frac{b}{c}.$$

Squaring both sides and summing leads to

$$\sin^2\theta + \cos^2\theta = \frac{a^2 + b^2}{c^2} = 1,$$

so that

$$\cos\theta = \sqrt{1 - \sin^2\theta}.$$

Newton must have wondered whether his theory was capable of accounting for the elliptical orbits of the planets. Kepler had given a detailed description of this motion, which will be discussed in Chapter 5. In an elliptical orbit, a planet's distance from the sun varies. It is necessary to know how the gravitational force varies with distance from the sun in order to account for this motion. Circular motion is a special case of the more general one. It is likely that Newton dealt with this case first. If a planet is in such an orbit about the sun, Kepler's laws tell us that it will be traveling at constant speed, describing what is known as uniform circular motion. Moreover, these laws tell us how the period (the time it takes to make an orbit) depends on the radius of the orbit. Much of this chapter will be devoted to studying this motion. Since the motion is confined to a plane, only two coordinates are needed for its description. It is convenient to pick the origin of the coordinate system at the sun's center, which is also the center of the circular orbit. Polar coordinates (r, θ), as shown in the figure below, can be used to describe this motion. These coordinates are simply related to Cartesian coordinates. From the diagram, it is seen that

$$x = r \cos \theta, \qquad y = r \sin \theta, \qquad x^2 + y^2 = r^2$$

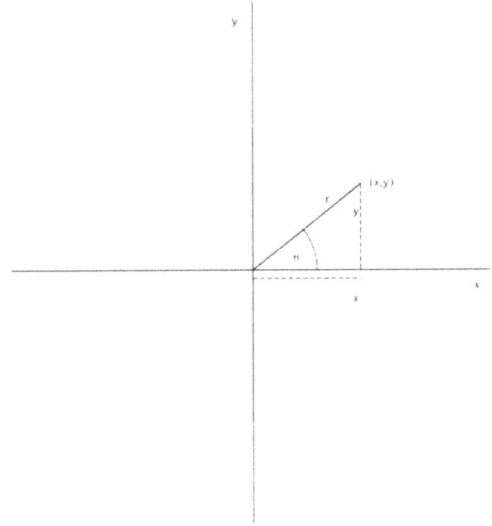

Note that the original definitions of the trigonometric functions, sine and cosine, were limited to angles smaller than 90°. Use of the equations that relate the two coordinate systems extends these definitions. Thus, when an angle is between 90° and 270°, the cosine becomes negative. This follows from the fact that the coordinate x is negative in this region. Note that r is always positive. The sine of an angle is negative for angles between 180° and 360° because the y coordinate is negative for this range of angles.

For circular motion centered about the origin the polar coordinate, r, is a constant. It is the radius of the orbit and is denoted by R. Assuming the x-axis is chosen to be in the direction of the initial position the angular coordinate is given by the equation

$$\theta = \omega t.$$

In this equation, ω (the Greek letter *omega*) is a constant. It is called the angular velocity. In the general case, in which the angular velocity is not necessarily constant, it is defined by the equation

$$\omega = \frac{d\theta}{dt}.$$

In uniform circular motion, both the position vector and the velocity vector of a particle are constant in magnitude. The magnitude of the position vector is the radius of the orbit, and the magnitude of the velocity vector is the speed of the particle. The velocity is the rate of change of position, while the acceleration is the rate of change of velocity. Both of these vectors are changing because their directions are changing. They both are proportional to the angular velocity. If the dependence of v on R can be found, the same analysis can be used to obtain the dependence of a on v. The dependence of the velocity on position is taken up first.

There are two aspects to the problem at hand. The first is to determine the speed. It is proportional to the angular velocity. The usual choice of units, such as revolutions per second or degrees per second, is not very convenient. For our purposes, it helps to choose an angular measure called the radian. Consider the figure below. The ratio of the arc length, S, to the radius, R, is proportional to the angle and depends on nothing else. The radian is chosen as the unit such that the constant of proportionality is one. In other words, the radian is the value of theta when S = R. Recall that the ratio of the circumference of a circle to its diameter is π (the Greek letter *pi*). Its value, to an accuracy of five decimal places is 3.14159. The diameter is double the radius, so that one revolution, or 360⁰, corresponds to 2π radians. One radian equals 57.2958⁰, to a very good approximation. The exact figure is 180/π degrees.

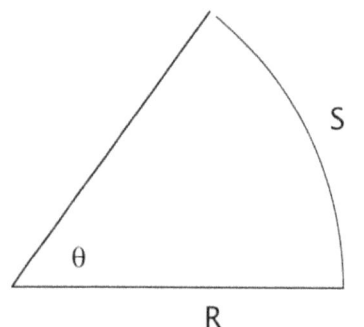

Consider a particle that is moving in a circular orbit, as in the adjacent figure. S could represent the distance it has traveled in some time, t. If the angle θ is measured in radians this distance is given by the equation S = Rθ. The derivative, dS/dt, is the speed, since it corresponds to the rate at which the distance is increasing. The formula for the speed is

$$v = \frac{dS}{dt} = R\frac{d\theta}{dt} = R\omega.$$

The direction of the velocity vector is perpendicular to that of the position vector, as measured from the origin. For example, if the particle were to the right of the origin in the figure above, it would be traveling vertically.

The dependence of the acceleration on velocity is completely analogous to the one that was just used for the dependence of velocity on position. It is a result of the velocity vector rotating with the same angular velocity as that of the position. The result for the magnitude is thus given by the formula

$$a = v\omega,$$

where v is the speed that was found in the previous equation. The direction of the acceleration is perpendicular to that of the velocity. You should be able to verify that it is directed opposite to the position vector. In other words, it is toward the origin.

Newton's discovery that a planet orbiting the sun in a circle is accelerating toward the sun must have been an Aha moment! He had found a way to define velocity and acceleration at any instant. He had generalized the meaning of the words *velocity* and *acceleration*. He had postulated a law of motion that was consistent with the motion of the planets and also fit Galileo's description of falling bodies. His assumption that the force that the sun exerted on a planet was gravitational was consistent with the idea that the motion didn't depend on the planet's mass. He must have had a powerful feeling that he was on the right track. For centuries, it had been believed that astronomical bodies were somehow different from anything found on earth. To state that they were merely chunks of matter that obeyed the same rules as the ones on Earth was unthinkable. Try to imagine the impact that Newton's ideas had on the astronomical community as well as that of the public, when these ideas became known. Of course, he still had some work to do. A circular orbit is a special case of the more general elliptical ones described by Kepler. Newton had to develop a theory of gravitation in order to do this. He didn't publish his theories until he had shown that elliptical orbits were consistent with his ideas. It was the astronomer Halley, the discoverer of the comet that bears his name, who convinced Newton to publish *The Principia*, a book that described all of his work.

According to Newton's third law, there must be a reaction force from a planet, back on the sun. Does the sun accelerate? It probably is a good approximation to treat the sun as being so massive that its acceleration is extremely small. Moreover, it is being pulled in many different directions by all the matter in the solar system. It is interesting to note that astronomers today are discovering planets in orbit around a star by the star's wobble. Newton would have understood this research. Although stars are not exactly stationary, it is an excellent approximation to treat the frame of reference in which the sun is stationary as an inertial frame. This frame does not rotate with respect to the stellar dome.

The speed and acceleration of a particle undergoing circular motion, with angular velocity ω and radius R, have been shown to be given by the equations

$$v = \omega R, \quad a = \omega v.$$

The first equation can be used to eliminate either the speed or the angular velocity from the second equation. The acceleration can then be expressed in either of the following ways:

$$a = \omega^2 R, \qquad a = \frac{v^2}{R}.$$

These equations make sense. On a merry-go-round, everyone is going at the same angular velocity. Those who are farthest from the center experience the greatest acceleration. On the other hand, on a highway, most of the cars are traveling near the speed limit. In other words, they are probably all going at nearly the same speed. When rounding a curve, those cars near the center experience the greatest inward acceleration. If a passenger in a car doesn't get the required inward force from the seat upon which she sits, she doesn't participate in the car's motion and slips outward until the door provides the required force. It all makes sense. You should try to make sense out of your observations of circular motion without having to use the words *centrifugal* or *centripetal*. Can you explain why water need not spill out of a bucket that is being swung around in a vertical circle without using either of these words?

Can you use what you have learned so far to calculate the speed of a spaceship orbiting the earth, near the surface? In actual practice, it is necessary to get beyond the problem of atmospheric friction. Neglect the small effect that this requirement has on the radius of the orbit or on g. In other words, use g = 32ft/sec². This can be equated to v^2/R, where R is the radius of the earth (about 4,000 miles). Use of this formula leads to v = 17,700 mph.

Congratulate yourself if you question the assertions made in the last paragraph. The earth is not an inertial frame. It is necessary to correct for the fact that the earth is spinning. For another, the sun has been ignored. The earth is accelerating toward the sun. A frame fixed with respect to the earth wouldn't be inertial even if the earth weren't spinning. The sun's force on the spaceship of the previous paragraph has also been ignored. It turns out that the two effects associated with the sun almost cancel. This will be clarified later in the chapter, in the section called "The Tides."

Let's suppose the spaceship is launched at the equator, eastward. The ground there is already moving at a speed of about 1,000 mph with respect to the stellar dome. (A point on the equator travels roughly 25,000 miles in about 24 hours.) The speed of the spaceship with respect to the ground is about 16,700 mph. It should then take about 1.5 hours to make an orbit. This is in remarkably good agreement with the actual orbit times of low flying satellites. Newton didn't have these satellites to reassure him he had the right idea.

Newton's Law of Gravitation

In order to deal with elliptical planetary orbits, it is necessary to know how the gravitational force depends on the distance of a planet from the sun. Newton was aware of Kepler's laws. One of the laws had to do with the period of planetary orbits. When applied to a circular orbit, the law said that the square of the time to complete the orbit is proportional to the cube of the radius. Newton must have realized that he could use this law to obtain the result he needed. He had an expression for the acceleration in circular motion that depended on the angular velocity. The period of a circular orbit is inversely proportional to the angular velocity. It follows that the square of the angular velocity is inversely proportional to the cube of the radius of the orbit. This can be written as

$$\omega^2 = \frac{K}{r^3},$$

where K is a constant of proportionality. Newton could now make use of his second law as follows:

$$F = ma = m\omega^2 r = m\frac{K}{r^3}r = \frac{Km}{r^2}.$$

Newton realized that the gravitational force between two bodies would be proportional to the mass of each body, not just one of them. Thus, the constant, K, must involve M, the mass of the sun. The formula that emerged was

$$F = G\frac{Mm}{r^2},$$

where G is called the universal gravitational constant. It wasn't known in Newton's time. John Michell, a geologist, designed a delicate experiment for the purpose of measuring G some time before 1783. However, he died, and the experiment wasn't carried out until 1787 by the British scientist Henry Cavendish (1731–1810). It is recommended that you read about the experiment on the Internet. The numerical value of G in the MKS system is 6.74×10^{-11}.

Newton wasn't satisfied with this result, as it only applied to the sun and the planets. He wondered if all the particles in the universe attracted one another in accordance with such a law. He postulated this as a law of attraction between all particles. He was then able to verify by a complicated argument that a body that had spherical symmetry would behave as though all its mass were concentrated at its center. It is necessary to put in a cautionary statement. If a spherical body is hollow, it is necessary to say that it behaves like a point only for the region outside itself. Newton's universal law of gravitation, as the more general law is known, thus allows us to treat the sun and the planets as particles as far as their gravitational behavior is concerned. Newton was amazed by his own theory. The fact that a force could reach out through empty space was almost inconceivable in his time. He wrote to his contemporaries about his misgivings.

It is of interest to check Newton's theories with the orbit of the moon. The moon is roughly 60 times as far away from the earth's center as a point on the earth's surface. The square of 60 is 3,600. The acceleration due to the earth's gravity at that distance ought to be $32/3,600$ ft/sec². From this fact, it is found that

$$\omega^2 r = 32/3600.$$

The approximate figure of 240,000 times 5,280 is used for r. This equation is solved for ω, from which the period is found to be 27.5 days. Newton's theories seem to work!

The behavior of g is a surprising one. Why should the mass of a sphere behave as if it were concentrated at the geometrical center? Newton was able to prove that it actually did this by means of integration, a process by which he could sum the contributions of all the particles. However, a later proof by the great mathematician Frederic Gauss is much simpler. This method will be used again in dealing with electromagnetism. It is taken up at the end of this chapter. It is based on an ingenious idea that takes advantage of the symmetry of a sphere. Modern physicists make great use of symmetry ideas.

In calculating the motion of the moon around the earth, the force on the moon by the sun was ignored. Actually, the force that the sun exerts on the moon is more than twice the force that the earth exerts on it! This can be deduced from the following analysis: The moon's distance from the sun varies, but it is roughly the same distance as that of the earth, namely 93 million miles. If it were subject to no other forces other than the one from the sun, its acceleration toward the sun would be the same as that of the earth. All freely falling bodies have the same acceleration at a given location. The ratio of the sun's force on the moon to that of the earth's force on it is the same as the ratio of the two accelerations that were just discussed. The formula $\omega^2 R$ can be used for the acceleration. First compare the angular speeds for the two orbits. There are approximately 13 lunar months in a year, which means 13 revolutions of the moon about the earth for one revolution of earth about the sun. The angular speeds are in the ratio of 13 to 1. When squared, the ratios are roughly 170 to 1. Now compare the radii of the orbits. That ratio is roughly 93 million to 1/4 million. In round numbers, it is 370. Thus, the acceleration toward the sun wins out by a factor of about 37/17.

Although the sun's force on the moon is more than twice as strong as that caused by the earth, it is a good approximation to ignore this force when describing the motion of the moon about the earth. This is explained in the next section.

The Tides

Physicists often talk of freely falling elevators, considered as frames of reference. Imagine such an elevator, subject just to the force of the earth's gravitational pull. If you were standing in this elevator, you would accelerate downward at the same rate as the elevator, without necessarily making contact with the floor. Both you and the elevator would have the same acceleration, subject only to the gravitational force. If you were to throw a ball in such an elevator, it would also have the same acceleration relative to the earth. Its acceleration with respect to the elevator would be zero. It is similar to the fact that the relative velocity of two cars is zero when their velocities are the same. In this elevator, you could reasonably claim that there was no gravity. A spaceship, in orbit about the earth, is equivalent to the freely falling elevator, in that it is coasting under the influence of a gravitational force. If the ship is not rotating with respect to the celestial dome, it also behaves like an inertial reference frame. A passenger could claim that the gravitational field of the earth is absent inside this spaceship, since she cannot discern any effects of the earth inside this ship.

Instead of elevators and spaceships, consider the planets. They are also moving under the influence of gravitational forces, primarily from the sun. As far as motion with respect to our planet is concerned, the sun's gravitational force should be ignorable. The only difference between the earth and the spaceship is that the earth creates a measurable gravitational force field of its own, while the spaceship's gravitational field is negligible. The moon's motion about the earth can be regarded as being due to the earth's gravitational pull. The sun's pull on it is stronger, but as far as motion about the earth is concerned, the earth's acceleration toward the sun cancels out the effect of this. Tidal forces are being ignored here. It will become clear what this means in the next few paragraphs.

Consider a futuristic spaceship orbiting the earth that is hundreds of miles in diameter. The gravitational pull by the earth on a particle moving about within it would vary noticeably as the particle moves about. When it is nearest the earth, the particle feels a stronger downward pull than it does on average. It also feels a weaker-than-average pull near the top (the part farthest from Earth). The particle always feels a downward pull, but it already has been shown that *down* to one observer near the earth isn't necessarily in the same direction as *down* to another. The word really should mean "toward the earth's center," implying that lines pointing downward are not parallel. They are converging because they intersect at the earth's center.

What are the consequences of the gravitational force varying in strength and direction? Imagine that a huge ball of ice were placed near the center of this futuristic spaceship so that it floated in equilibrium. If the ice was then heated so that it started to melt at the surface, what would happen? Particles of water near the top would tend to fall less rapidly than the others, and they would tend to move upward. Meanwhile, particles near the bottom would tend to fall more rapidly than the average and would tend to move downward. The liquid sphere would distort so as to be stretched in the vertical direction. At the same time, the sides would be compressed because the downward forces converged.

There is no difference between the imaginary spaceship and the earth, except for the fact that the earth has a strong gravitational force of its own. Our oceans and atmosphere are affected like the water on the spaceship. The tidal forces are fictitious. They arise because the earth is not an inertial frame. The situation is more complicated than the example of the spaceship. For one thing, the sun and the moon are both involved. For another, the fact that the earth is rotating makes the tidal force time dependent at any given location on the earth. Lastly, the earth has geographical features that affect the motion of the water. There is a huge response in Canada's Bay of Fundy, for example, because of a type of resonant behavior. From the description of the nature of tidal forces, two high tides per day are expected. This is actually what is observed.

Why does the moon's tidal effect compare with that of the sun? You have already read that the sun's force on the moon is twice as great as that of the earth. The earth is about 80 times as massive as the moon. The sun's force on the earth is, thus, about 80 times the sun's force on the moon. The sun's force on the earth is over 160 times as great as the earth's force on the moon. The sun's force on the earth is therefore over 160 times the force of the moon on the earth. If the last sentence is unclear, recall Newton's third law. You might think that the moon's effect on the tides can be ignored in comparison with that of the sun.

So, why is the moon responsible for larger tidal forces than the sun? A qualitative answer is given. What counts, as far as the tidal forces on the earth are concerned, is not the size of the force on a particle, but how that force changes when the particle goes from one side of the earth to another. The moon is 250,000 miles away, and the earth's diameter is 8,000miles. The distance to the moon changes by a few percentage points when going from the near side of the earth to the far side. As for the sun, the 8,000-mile difference in distance makes less than a .01 percent difference. Recall that the distance to the sun is about 93,000,000 miles. The larger change in force due to the moon's proximity to the earth is ultimately responsible for its dominance in the matter of tides.

The fact that both the sun and the moon are significantly involved in the tides makes it clear that the tides will be largest when they are in harmony, rather than when they are opposing one another. It is a good idea to check the calendar before visiting the Bay of Fundy in order to witness the huge tides. It may be safer to read about the predicted behavior of the tides on the Internet.

Thought question: When is the moon most in harmony with the sun? When it is full, when it is new, at half moon? When is it least in harmony?

Before leaving the subject of tidal forces, it is interesting to consider the gravitational effects of black holes. These are objects that emit no light or other radiation. They are so massive that the gravitational forces have caused them to collapse into what is essentially a point. Actually, our current knowledge of physics is insufficient to tell us much about this singularity. It probably isn't actually a point. Once inside a boundary known as the event horizon of a black hole, nothing can escape. Although black holes can't be observed directly, their effects on neighboring bodies can be profound. They were predicted to exist as a consequence of Einstein's general theory of relativity. They can be so destructive that Einstein didn't think that nature would ever allow them to come into being. Once a body falls into a black hole, the tidal forces get bigger and bigger. If a person were to fall into one feet first, the pull on the feet would get to be enormously greater than the pull on the head. The fate of the individual would be horrible under such a stretch. The forces are expected to pull apart everything that falls into a black hole, even atoms. There is some controversy of this supposed fate. Some theorists claim that space and time end at the event horizon. There will be more on this subject later.

Simple Harmonic Motion

Although Newton's initial claim to fame came from his explanation of all of Kepler's laws, it is a good idea to postpone the discussion of these laws until the theory has been applied to a few other phenomena. This will lead to a deeper understanding of the meaning of the laws. Concepts will be developed that will prove helpful in dealing with elliptical orbits. It will be assumed, unless explicitly stated otherwise, that Newton's laws are valid on the earth. It is an excellent approximation for many phenomena.

Newton's second law is applied to a vibratory motion known as simple harmonic motion, or SHM. This motion can be visualized as uniform circular motion viewed from the side. Since the motion is one-dimensional, only one Cartesian coordinate is needed. The radius of the circular motion associated with SHM determines the amplitude of the vibration. For this reason, the radius is denoted by the symbol A, and the equations for the circular motion in polar coordinates are

$$r = A, \quad \theta = \omega t + \theta_0.$$

Note that the angle need not be 0 when the time variable is 0. This allows for the treatment of two different SHMs that are not in phase with one another. This topic will be dealt with in a later chapter. In Cartesian coordinates, the equations for the circular motion are

$$x = A\cos(\omega t + \theta_0), \quad y = A\sin(\omega t + \theta_0).$$

It is only necessary to set either one of the two Cartesian variables to 0 in order to describe SHM. Since the choice is arbitrary, it will only be necessary to focus attention on the first of these two equations. Thus, the only Cartesian coordinate that is changing with time is x.

Consider the following question: "What kind of force could produce such motion?" In order to answer this question, it is first necessary to determine the acceleration. Newton's second law will then determine the force.

Those readers who are familiar with differential calculus can differentiate x with respect to time to get the velocity. A second time derivative leads to the acceleration. There is another way that avoids this knowledge of mathematics. It is only necessary to make use of the associated circular motion and take the x-component of all the vectors concerned.

The diagram below is a schematic that shows the relative directions of the position, velocity, and acceleration vectors for the circular motion at some instant of time. Note that the sizes of the vectors can't be compared since they have different units. It makes no sense to compare speed with distance, for example. In order to find the position, velocity, or acceleration of the SHM at this instant of time, it is only necessary to get the x-component of the corresponding vector. In order to get this, just imagine that the vectors shown are considered to be the hypotenuse of an appropriate triangle. You should be able to obtain the equations shown below the figure.

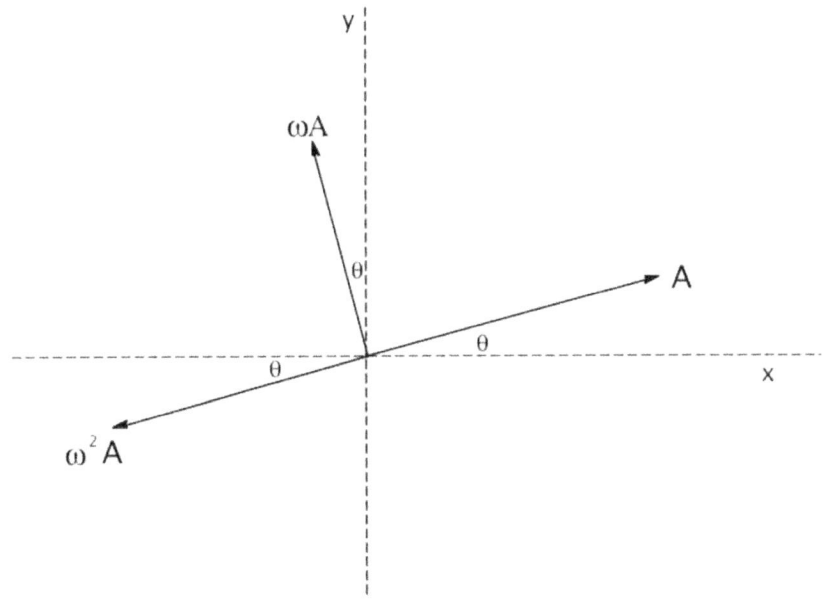

$$x = A\cos(\omega t + \theta_0),$$
$$v = -\omega A\sin(\omega t + \theta_0),$$
$$a = -\omega^2 A\cos(\omega t + \theta_0).$$

Note: In a later chapter when elliptical planetary orbits are considered, it will be necessary to use the following equation:

$$\frac{d\cos\theta}{d\theta} = -\sin\theta.$$

This equation can be derived from the fact that v is the time derivative of x , If this point is not clear, it helps to note that that the name of the variable doesn't matter. In other words, theta could be replaced by t, and omega can be assumed to be one radian per second.

In a similar way it can be noted that

$$\frac{d\sin\theta}{d\theta} = \cos\theta.$$

This follows from the fact that the acceleration is the time derivative of the velocity.

Comparison of the third equation with the first reveals that

$$a = -\omega^2 x.$$

This equation is significant. It tells us that the acceleration is proportional to x and in the opposite direction. The simplest case of SHM is associated with the oscillation of a body of mass, m, hanging from a spring. For reasonably small stretches or compressions, springs behave elastically, which means that the force that it exerts on the body is proportional to the stretch. This means that if a body of weight, W, causes a stretch, d, of the spring when hanging still, it follows that the stretch will be 2d when the body is replaced by a body of weight 2W. The net force on the body is given by the formula

$$F = -kx..$$

This is Hooke's law that states that the stretch of a spring is proportional to the force applied. This same equation can be used to specify the net force on a body hanging from a spring if x is measured from the equilibrium position. This is the position when the weight force cancels the force of the spring on the body. The constant, k, is called the spring constant. The negative sign indicates that the force is in the opposite direction from that of x. Newton's second law leads to

$$a = \frac{F}{m} = -\frac{k}{m} x.$$

This is consistent with the expression for a in SHM if

$$\omega = \sqrt{k/m}.$$

The frequency of oscillation, f, corresponds to the number of vibrations per second. It is the same as the angular velocity, if the latter were given in revolutions per second. Since the angle has been measured in radians, the frequency is

$$f = \frac{1}{2\pi} \sqrt{k/m}.$$

The factor 2π comes from the fact that there are 2π radians in one revolution.

Strictly speaking, Hooke's law applies only to springs that are in equilibrium. It relates the tension in a spring to the stretch. Imagine a spring that is stretched in a region where there is no gravity. If it is in equilibrium, the net force on it is zero. For this reason, the force at one end must cancel the force at the other. Thus, the forces are of the same size although in opposite directions. However, when the center of mass of a spring is accelerating, the size of the force at the two ends must differ. It is then not possible to talk about the tension in the spring as though it were a single number.

The formula for SHM is clearly not rigorous, because it implies that a spring would oscillate with infinite frequency if there were no body attached to it. However, the formula becomes more exact if the mass of the spring is very small in comparison with m. A rigorous treatment for a massive spring would show that SHM is correct but that the formula for the frequency involves the mass of the spring as well as the mass of the object suspended from it.

The particular solution x = Acos(ωt), for which θ_0 = 0, corresponds to the situation in which the body starts from rest. This can be seen from the equation for v (v= -A sin(ωt)), since the sine of a 0^0 angle is 0. The graph below is a plot of cos θ versus θ, using radian measure for the angle. You can visualize the sine function by shifting the curve $\pi/2$ radians to the right.

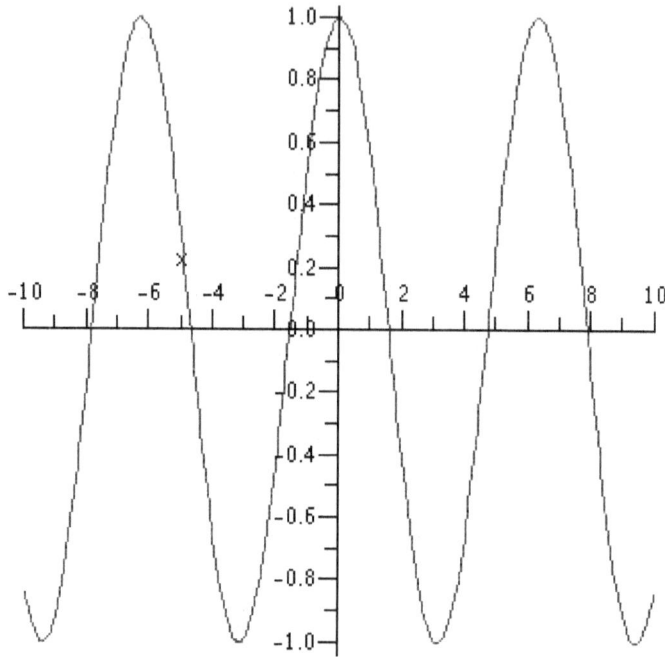

The type of oscillatory motion just considered is called sinusoidal regardless of the value of θ_0. Sinusoidal oscillations occur almost everywhere in nature. A tuning fork produces such an oscillation. A high frequency corresponds to a high pitch. The sound of a single note of music from a musical instrument is made up of a superposition of sinusoidal vibration of various frequencies. The lowest of these frequencies is called the fundamental and characterizes the note. The other frequencies are called harmonics. Each harmonic has a frequency that is an integer multiple of the fundamental. The difference between the sound of a trumpet and that of a clarinet comes from the difference in the strengths of their harmonics.

The Simple Pendulum

The oscillation of a pendulum approximates SHM when the amplitude is small. A simple pendulum consists of a weight hanging from a string of negligible mass. Such a pendulum is illustrated in the figure below. Let the length of the string be L, and the mass of the hanging body be m. More precisely, L corresponds to the distance between the center of mass of the object to the pivot. The center of mass then swings in a circle of radius L. The Cartesian coordinates of the mass are measured from the equilibrium position corresponding to $\theta = 0$. Recall that the weight, W, equals mg.

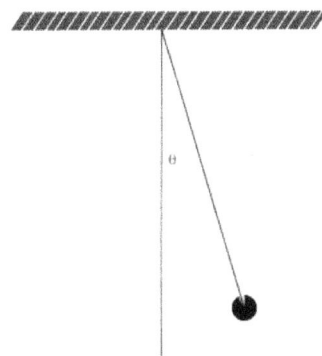

The horizontal position, x, of the weight is given by $x = L \sin \theta$. Consider the forces acting on the bob (the object at the end of the string). They are shown in the figure below. Such a figure is known as a free body diagram. It uses arrows and algebraic symbols to express what is known about all the forces acting on the body of interest. It is often helpful to have such a diagram when applying Newton's second law. In the present case there are only two major forces. Air resistance and buoyancy are ignored here as being too small to make much difference. The relevant forces are the weight, W, and the tension, T, in the string, at the point of contact with the bob.

If you consult the diagram, you can readily see that Newton's second law leads to two equations. The one that is associated with the vertical motion is

$$T\cos\theta - W = ma_y,$$

where a_y is the vertical component of the acceleration of the weight.

For small enough amplitude, a negligible error is made by setting the right side of this equation to 0, and by assuming $\cos\theta = 1$. The approximation amounts to the statement that

$$T = W = mg.$$

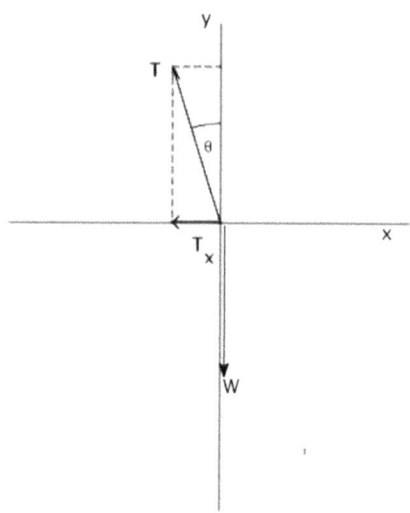

The second equation is

$$-T\sin\theta = ma_x.$$

The negative sign stems from the fact that the positive direction for x is to the right. From the original diagram, it can be seen that

$$\sin\theta = \frac{x}{L}.$$

The second equation simplifies to

$$-mg\left(\frac{x}{L}\right) = ma_x$$

$$a_x = -\frac{g}{L}x.$$

This equation results in SHM, since the acceleration is proportional to the displacement and is in the proper direction. Comparison with the earlier equations for SHM yields the following:

$$\omega = 2\pi f = \sqrt{\frac{g}{L}}.$$

In this equation, f is the frequency, or the number of vibrations per second. The reciprocal of f is the period, or the time it takes to make one oscillation. Unfortunately, the common symbol for this time of oscillation is T, the same symbol that was used for tension. Hopefully you won't be confused by the following equation,

$$T = 2\pi\sqrt{L / g},$$

for the period of a simple pendulum undergoing small oscillations.

The period of a pendulum is simple to measure. An accurate value can be obtained by allowing it to swing through hundreds of vibrations and measuring the time it takes to do so. It is easy to see how accurate values of g can be determined, without the necessity of timing the flight of an object in free fall. Although the theory that has been used here deals with a simple pendulum, it can be extended to deal with more complicated ones made out of rigid materials. It will be necessary to add one more idea to Newton's laws before that can be done. Rigid bodies will be taken up in Chapter 7.

Time

It was Galileo who first noticed that the time of swing of a pendulum didn't depend on the amplitude of the swing (as long as the amplitude was small). His son was the first person to make use of this fact in the construction of a clock. Although clocks had been around for several centuries, they were very expensive and erratic before the introduction of pendulums. How could scientists or clockmakers calibrate their clocks? The development of accurate clocks is a kind of bootstrap operation. It is necessary to find some periodic phenomenon to choose as a standard. You might think that the second could be defined in such a way that there are exactly

3600 seconds in an hour and 24 hours in a day. It would only be necessary to count the number of vibrations that a pendulum made in the course of a day in order to get an accurate measure of its period. It would be necessary to determine the precise moment when the sun made a transit, in order to be sure that a full day had elapsed. Even if an astronomer could do this, it would be found that the number of oscillations varied from one day to the next. The length of the solar day varies considerably over the course of a year. This will be explained later in this section.

A better way to determine the period of a pendulum before accurate clocks had been built is to make use of the sidereal day. This is the length of time it takes for the astronomical dome to make one revolution. It is only necessary to observe the transits of a single star past a particular longitude in order to use this for calibrating a pendulum. The number of seconds in a sidereal day is easily determined. There is exactly one more sidereal day in the course of a year than there are solar days. If it is assumed that there are 365.25 days in an average year, the number of seconds in a sidereal day is given by the following:

$$24(3600)\frac{365.25}{366.25} = 86,164.$$

In the world of physics and astronomy, it is necessary to check standards as the need for precision becomes greater. Standards change when better ones are found. Although the second had been based on the earth's motion for several centuries, this is no longer the case. There is good evidence that the angular velocity of the earth is slowing. Today, there are clocks that are accurate to 1 second in 300 million years. The second has been redefined so as to be a particular fraction of the vibrations associated with Strontium atoms held under extremely controlled conditions. The theory and the tools of physics are subject to continuous refinement. By the time this book is written, the standards may have changed again.

It is interesting to examine why the solar day varies in length. Those readers who keep track of the time of sunrise and sunset may have observed some of the effects of this variation. If you live near Albany, New York, for example, you might note that the earliest sunset of the year takes place around December 8. This may surprise you because the shortest day (sunrise to sunset) takes place on the solstice (around December 21). The latest sunrise occurs around January 8. The cause of this surprising behavior is that the sun is taking longer than average to complete its cycle at this time of the year. In other words, the length of the solar day is longer than average. Thus, even though the interval between sunrise and sunset is getting shorter in early December, it is possible for the sunset to occur later from one day to the next.

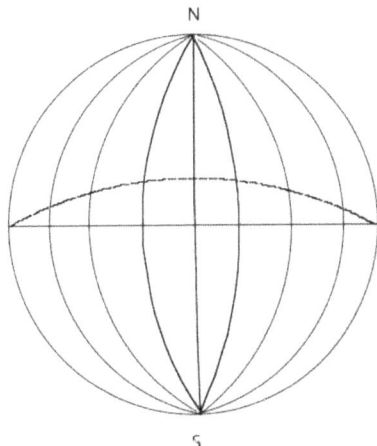

To make sense of this behavior, consider the figure above. It shows the path of the sun with respect to the earth that would occur in the course of a year if the earth weren't spinning. The horizontal line represents the equator. Lines of longitude are also shown. The path of the sun is shown by the dotted line. To make this clear, the sun is treated as a point located at its geometric center. Like any point in space, its location with respect to the earth can be specified by latitude, longitude, and altitude. The path that is shown ignores the altitude. The exact location of the sun at any instant depends on the precise moment when the earth's spin had been interrupted. It is the motion of the sun, not the position that is of concern here. If the earth's orbit were circular, the sun would be moving along this dotted line at constant speed. It shouldn't take too long to figure out that the motion is to the right, or easterly. Its angular speed is roughly 1^0 per day, as it takes a year to make a revolution. Note that the rate at which the sun crosses meridians varies for two reasons: 1. When the path is near the equator, the sun is traveling northeasterly or southeasterly rather than due east. 2. The meridians get closer together the farther one gets from the equator. The equinoxes occur when the sun's path crosses the equator. The solstices occur at the northern or southern extremes. In summary, the result of this analysis is that the easterly motion of the sun is greatest at the solstices and least at the equinoxes.

This fictitious situation of an earth with no spin can be related to the actual situation. The spinning earth causes the apparent motion of the sun to be from east to west. The easterly motion of the sun that was just considered has the effect of slowing the westerly motion caused by the spin. The slowing is greatest at the solstices and least at the equinoxes. Thus, the length of the solar day is longer at the solstices than at the equinoxes. The earth's orbit is not exactly circular. The earth is closest to the sun in December. The lengthening of the day is slightly greater in December than in June. It has a greater orbital angular velocity then.

Consider how lucky earthlings are that the earth is actually spinning. If not, there would be some locations on the earth for which the sun would be almost overhead for weeks at a time. Later on in the year, the sun would set and there would be almost 26 weeks of darkness. There would be certain regions for which the sun would almost always be near the horizon. As will be seen later, there would also be no wind at all. It is hard to imagine how life could exist on such a planet.

Although the discussion has centered on the measurement of time, time itself is a concept that is usually taken for granted. Although there are ways of measuring it, the word *time* is hard to define. Suppose you think of yourself as an actor in a movie, and the projectionist starts playing the film faster and faster. You would move faster, speak faster, understand faster, and react faster. Nothing at all would seem different to you. Only those people watching the movie would notice any changes.

It might surprise you to learn what physicists today are telling us, namely that the projectionist could play the movie backward without the movie audience seeing any physical laws being broken. Consider what you have learned so far. If you watched a movie of a mass on a spring oscillating up and down, or a pendulum swinging back and forth, could you tell if the movie were being played forward or backward? No law of nature would be broken if the planets were moving and spinning in the opposite direction.

You probably can think of phenomena in which you can be sure the movie is being shown backward. A classic example is one in which someone dives into a pool, making a big splash. When the movie is shown backward, the splash disappears, and a person emerges, flies upward, feet first, miraculously ending up on the diving board, dry and neat, while the water in the pool is nice and calm. Impossible? No. Improbable? Very. So improbable that you can bet any amount of money, and give any odds, that it never happened. Looked at microscopically in full detail, no laws of physics have been broken in the backwards motion. But you have to wonder how all those molecules that make up the splash could have arranged themselves in just such a way as to form a pressure gradient in the water sufficient to launch the person back up onto the board. This example is just one of many. All those phenomena that look irreversible have one feature in common. The laws governing the behavior of the individual molecules are not being disobeyed under time reversal. The irreversibility is connected to probabilities. The subject of *time* comes up again and again. It will be taken up in later chapters. Some scientists even wonder whether time is some sort of illusion.

Gauss's Derivation of Gravitational Law for Spheres

Consider the topic taken up earlier, of the gravitational field produced by a spherical mass distribution. The mathematician Friedrich Gauss (1777–1855) saw an interesting parallel between the gravitational field around a spherical planet and the velocity pattern of a fluid under certain conditions. In order to understand this, it is helpful to consider first the flow of water in a pipe. Treat the water as though it were incompressible and flowing in a pipe of cross-sectional area A. The volume of water, V, contained in a section of pipe that is of length L is the product AL. If the speed of the flow is v, the rate at which water would pass through would be Av. If lengths were measured in feet and time in seconds, the unit would be cubic feet per second. (This follows from the fact that each second a column of water of length v passes through the pipe.) If this pipe were just one section in a series of pipes, the speed, v, in each section would differ. The speed would be inversely proportional to the area. It is assumed that the speed of flow is constant within a given pipe. In other words, all the particles have the same velocity.

Now, imagine that all of space is filled with a huge sea of water. At first all is still. Then a kind of black hole is inserted in this infinite ocean. It is a particle that eats up water at a constant rate of N cubic feet per second. This sets up an inward flow of water toward this particle. The velocity of each water molecule will be directed straight in toward this particle. Now ask yourself how the speed of the water will vary with distance. By means of symmetry arguments, you will conclude that all water molecules at a given distance, r, from the "water eater" will be traveling at the same speed. From the discussion of pipes, the conclusion will be that the speed is inversely proportional to the area of a sphere of radius r. The area of such a sphere is $4\pi r^2$, so that

$$v = \frac{N}{4\pi r^2}.$$

The fact that this velocity pattern is inward and inversely proportional to the square of the distance from the particle made

Gauss see the connection to the gravitational force pattern created by a particle. The analogy was perfect. Gauss was able to use this analogy to deduce the flow pattern that would be created by a spherical distribution of such "water eaters." As long as the flow outside the region occupied by these particles is being considered, the velocity only depends on the number of cubic feet of water consumed per second. The symmetry of the flow is the same as before, and nothing else matters. From the analogy with gravity, Gauss could conclude that a spherical planet has exactly the same gravitational behavior as a particle with the same mass at the center of the planet.

Note how Gauss's treatment of fluids enables you to calculate the gravitational field inside the earth. It is necessary to know how the density of the earth varies with depth. The fluid analogy makes it clear that the only matter that counts is the matter inside a sphere of radius r. Let R denote the radius of the earth. If r is greater than R, then all the mass is involved. If r is less than R, only a fraction of the mass of the earth is involved.

Ask yourself how g would depend on r when r is less than R if the density of the earth were a constant. There are two competing contributions. The fact that the acceleration is inversely proportional to the square of the distance from the center gives an increase by a factor R^2 / r^2. The mass within a sphere of constant density is proportional to the volume, which, in turn, is proportional to the cube of the radius. This reduces the acceleration by a factor r^3 / R^3. The end result is that the acceleration due to gravity in such a case is given by

$$a = gr / R.$$

The symbol g in this equation is the acceleration due to gravity at the earth's surface.

The above type of force field makes for an interesting result. Consider a hole made from the North Pole to the South Pole of such a planet. What kind of motion would an object have if dropped into such a hole? Neglect air resistance or friction. The answer is SHM about the center of the planet. The acceleration is downward and proportional to the distance from the center. The angular frequency, ω, is given by $\omega = \sqrt{g/R}$. It may be interesting to note that the time it takes for a package to go from the North Pole to the South Pole, according to this prescription, is the same as it takes an orbiting satellite to go halfway around the earth if it is near the surface of the earth, about 45 minutes.

Chapter 3
Momentum

There are often several ways of looking at a single idea. Newton's second law has been stated symbolically as **F** = m**a**. (Note that vector quantities are being represented by boldfaced letters.) This equation can be written as

$$\mathbf{F} = m\frac{d\mathbf{v}}{dt}, \text{ or alternatively as } \mathbf{F} = \frac{d(m\mathbf{v})}{dt},$$

because the mass of a particle is a constant. The product, m**v**, is called the momentum of the particle. The second law can then be stated as follows:

"The rate of change of momentum of a particle equals the net force on the particle."

This is surprisingly simple. It turns out to be surprisingly useful. Suppose two particles are in outer space just coasting along at constant velocity, free from all forces. They are heading for a collision with one another. When they finally do collide, they exert forces on one another. Whether they stick together or fly apart after the collision, the following observation can be made: The total momentum the particles had before the collision is exactly the same as the total momentum after the collision. In order to prove this, it should be noted that momentum is a vector. The direction of the momentum vector is the same as that of the velocity. When dealing with vectors, it's usually easiest to deal with one Cartesian component at a time.

Recall Newton's third law. When two bodies interact, the force on one is equal to and opposite the force on the other. One body may receive a force that is positive in the x direction. The other body receives an equal magnitude force in the negative x direction. The rate at which one particle's momentum in the x direction is going up is the same as the rate at which the other particle's momentum is going down. It could be said that the momentum in the x direction is being transferred from one particle to the other but that the total momentum shared by the two wasn't changed by the collision.

What was true for the x-component of momentum is true for the other Cartesian components. Each component is independent of the other. Thus, the total momentum is conserved. The word *collision* does not necessarily mean that the two particles make contact with one another. The interaction could have been gravitational. It doesn't matter.

This idea can be extended to any number of particles. It will be illustrated for the case of three particles as in the figure below. These particles may be interacting with one another but subject to no other forces.

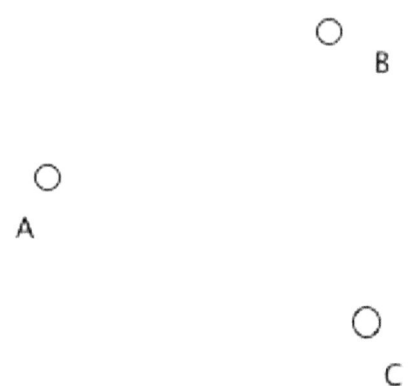

There are two forces on each body, so there are six forces in all. Assume that F_{BA} denotes the force that B exerts on A with similar notation for all the other forces. Note that F_{AB} is the reaction to F_{BA}. Newton's third law implies then that

$$F_{AB} + F_{BA} = 0. \; F_{AC} + F_{CA} = 0, \; F_{BC} + F_{CB} = 0.$$

The rate of change of momentum of body A is just the force on it, which is the vector sum of all the forces on it.

Rate of change of momentum of particle A = $\mathbf{F}_{BA} + \mathbf{F}_{CA}$
Rate of change of momentum of particle B = $\mathbf{F}_{AB} + \mathbf{F}_{CB}$
Rate of change of momentum of particle C = $\mathbf{F}_{AC} + \mathbf{F}_{BC}$

If these equations are added to one another, the sum of the terms on the LHS (left-hand side) equals the rate of change of momentum of the entire system. The sum of the terms on the RHS is 0. The conclusion is that the total momentum isn't changing. In other words, the total momentum of an isolated system is constant. This is a famous consequence of Newton's laws. That law still holds, over three centuries later, although it is now known that light carries momentum and that the formula, m$\mathbf{v}$, for the momentum of a particle is not quite right.

The statement is often made that momentum is conserved in a collision, even when there are external forces acting on the bodies involved. It is not a rigorous statement that holds for all collisions. However, it is often an excellent approximation. Consider the situation when one billiard ball collides with another on a pool table. The encounter is so brief that any frictional force that they experience from the table has a negligible effect in this time. If the external forces are weak compared to the ones involved in the collision, and the collision is sudden, this approximation can be justified, as though it were a law of physics.

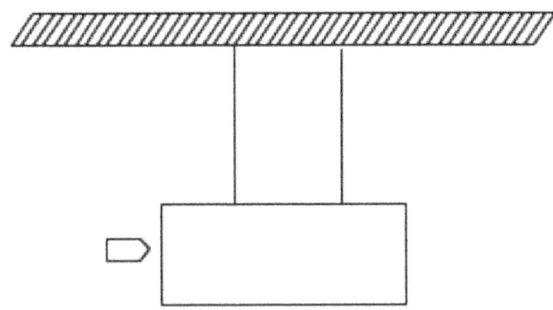

The diagram above is that of a ballistic pendulum. This is a device that was used to measure the speed of a bullet before high-speed electronic timing devices were invented. It makes use of the idea of momentum conservation. A bullet is fired horizontally into a chamber that is filled with sand and suspended by strings as shown. The chamber is sufficiently large so that the bullet remains embedded in the sand. The strings remain almost vertical during the time the bullet is slowing down so that momentum conservation is an excellent approximation during this time. The masses of the bodies involved in the collision are easily measured. It is only necessary to find the speed of the chamber right after impact in order to determine the speed of the bullet just before impact. The question of how to measure the speed of the chamber right after impact is taken up in the next chapter.

In the example of the ballistic pendulum, two objects combine to form a single entity. Momentum conservation also applies to the case where a single object bursts into pieces, such as in an explosion, or a rupture. It accounts for the recoil of a cannon or a gun. It also accounts for the propulsion of a spaceship by means of rocketry. Can you think of a use for the concept of momentum conservation as it relates to the following problem that an astronaut might face? Suppose she is making extravehicular repairs on her spaceship when her tether comes loose, and she starts to drift away from the ship. She has a bag of tools but no rockets. Is she helpless?

The concepts developed here can be used to calculate the thrust of a jet engine that is exhausting burned fuel with an average speed, v, measured in meters per second. Note that this speed is measured with respect to the engine. Let the rate at which mass is being ejected be given by the symbol R, in kilograms per second. In a reference frame in which the engine is at rest, the exhaust is gaining momentum at the rate Rv. An equal and opposite momentum is being imparted to the engine in order for momentum to be conserved. The conclusion is that the force on the engine, namely the thrust, is Rv Newtons. This follows from the fact that the force equals the rate of change of momentum.

In a similar manner, the force on the wall of a building from the impact of the water from a fire-hose can be calculated. It is the reverse of the previous situation in that there is an incoming flow, rather than an outgoing one. In order to make the calculation, it is necessary to make an assumption. Does the water stop when it hits the wall, or does it bounce? Note that the momentum transfer is larger if the water bounces.

Center of Mass

The center of mass of an object, or of a cluster of particles, is, in some sense, an average position. The word *average* has been used several times before. It is worthy of a definition. The average speed for a trip of 180 miles that lasts 4 hours is 45 mph. This means that if the speed had remained constant at 45 mph, the same distance would have been covered in the same time. When the class average on an exam is 80, it means that the total of all the grades would be the same if everyone in the class received a score of 80. If you wanted to get an average of grades on an exam for an entire school, in which you knew the average grade in each room, it would be necessary to make use of a weighted average. You would take the average in each room, then multiply that number by the number of students in that room, before summing in order to get the total score. The number of students per room could be called a weighting factor.

The center of mass is a weighted average of position in which mass is used as the weighting factor. Restrict your attention to one dimension first, and specify the position by the coordinate x. Particle #1, with mass m_1 is located at position x_1, and similarly for all the other particles. The weighted sum would then be

$$m_1 x_1 + m_2 x_2 + m_3 x_3 + ...$$

Let M stand for the total mass (the sum of all the masses). The x-coordinate of the center of mass, X, is obtained by dividing the weighted sum by the total mass or, equivalently,

$$MX = m_1 x_1 + m_2 x_2 + m_3 x_3 + ...$$

If the particles are moving, each coordinate depends on the time. If the time derivative of this equation is taken, the LHS is the total mass times the velocity of the center of mass. The RHS is the sum of all the momenta (rigorously speaking, the x-component of the momenta).

The following rule is worth remembering:

The mass of any object times the velocity of its center of mass equals the total momentum of the object.

The symbol that physicists commonly use for momentum is the letter **p**. It is shown as boldfaced because momentum is a vector. A capital letter is used to represent the momentum of the entire system. The rule is then written as

$$\mathbf{P} = M\mathbf{v}_{cm}.$$

As a consequence, the velocity of the center of mass of an isolated system is a constant.

It is no longer necessary to restrict your attention to one-dimensional systems. It is a simple extension of what has been shown so far to include external forces on a system. In that case, the rate of change of momentum of the system equals the net external force on the system,

$$\mathbf{F} = \frac{d(M\mathbf{v}_{cm})}{dt} = M\mathbf{a}_{cm}.$$

The external force is calculated by adding up the forces on the particles exerted by objects outside of the system being considered.

The law of the conservation of momentum was based on Newton's third law. It is of interest to discuss where these laws stand today in view of the fact that Newton's laws have been replaced by more modern theories. The law of the conservation of momentum still stands in relativity theory, although the equation for momentum has been altered. Newton's third law is not strictly valid because the concept of action-at-a-distance has been discarded. Let us make this point a bit clearer by an example. Suppose the position of the sun were to change suddenly. According to Newton, the force on the earth would change at the same time. In Einstein's relativity theory, the earth cannot obtain the information that the sun had moved until later. The delay is associated with the time it takes a light signal or a gravity wave to travel from the sun to the earth, a matter of several minutes. The interaction between the earth and the sun is delayed. This is in contradiction to Newton's ideas. Instead of a direct interaction between particles, it is necessary to imagine something that can propagate from one body to another. Thus, if the momentum of the sun should change suddenly, the momentum of the rest of the solar system wouldn't change at the same time. Today physicists can only claim that momentum conservation is valid by assuming that the gravitational "field" can carry momentum as well as material bodies. In the same way, electric forces can be propagated by electromagnetic waves that carry momentum. When you turn on a flashlight, there is a recoil, just as there is when you fire a gun. However, in the case of the flashlight, it is not likely that you will notice it! The momentum carried by a light wave can be measured by delicate instruments.

Answer to astronaut question: If the astronaut throws a tool in a direction away from the spaceship, she should acquire some momentum, propelling her toward the spaceship.

Chapter 4
Energy

In today's world, the concept of energy is almost intuitive. It is associated with electricity, with food and fuels, with motion and activity. These ideas developed relatively recently, primarily in the nineteenth century. In Newton's time, the concept of energy conservation was a million miles away, so to speak. Nevertheless, the concept of energy had its origins in Newton's laws. There is a superficial resemblance between energy and momentum. This can be seen by considering motion in one dimension. The force on a particle equals its rate of change of momentum with time. The energy can be defined for this case such that its rate of change with position is the force on the particle. Before going into details, it is a good idea to discuss the similarities and the differences between energy and momentum. It will be found that the energy of a particle depends only on its mass and speed. It is a scalar. This statement is true in a limited sense. It is not a scalar like mass, because the speed of a particle is different as seen from frames of reference that are moving with respect to one another. However, in a given reference frame, it is independent of the direction of the coordinate axes.

How is the speed of a particle related to the net force on it? Recall that acceleration is the rate of change of velocity, not speed. In the case of a particle executing uniform circular motion, its speed is constant even though it is subject to a force. In general, only the component of the force in the direction of the velocity has an effect on its speed. A perpendicular component of force only affects the direction of the velocity vector, not its magnitude. Since energy is a scalar, it is not concerned with the direction of the motion.

Forces that depend explicitly on the time are not encountered much in nature. What is meant by *explicitly* can be understood from the examples of circular motion and SHM already considered. In each case, the force changes with time. However, these forces really depend only on the position of the body. They change with time only because the position of the body changes with time.

Motion of a particle along the x-axis will be considered first. Consider its acceleration. It is defined as a time derivative of its velocity. If the velocity is considered to be a function of position, the time derivative can be written as

$$\frac{dv}{dt} = \frac{dv}{dx}\frac{dx}{dt}.$$

This property of derivatives is shared with that of ordinary fractions.

By making use of the definition of velocity for the one-dimensional case, the equation becomes

$$\frac{dv}{dt} = v\frac{dv}{dx}.$$

This can be written as

$$a = \frac{dv}{dt} = \frac{d(v^2/2)}{dx}.$$

In order to show this, the fractional property of derivatives is exploited once more. The quantity in parentheses is a function of v. It depends on x only because v depends on x. Thus, it is only necessary to differentiate the expression in parentheses with respect to v and then multiply by dv/dx. You have already seen that the time derivative of t^2 is 2t (for the motion of a falling body). In a similar fashion, the derivative of v^2 with respect to v is 2v.

It is now possible to write Newton's second law for the case of a particle in one-dimensional motion as

$$F = ma = \frac{d(mv)}{dt} = \frac{d(\frac{1}{2}mv^2)}{dx}.$$

Although this equation has been derived for a special type of one-dimensional motion in which the force depends only on position, it turns out that the quantity in parentheses is worth naming. It is now called the kinetic energy of the particle. The word *kinetic* implies that this energy is due to the particle's motion. The formula is written as

$$K = \frac{1}{2}mv^2,$$

so that Newton's second law becomes

$$\frac{dK}{dx} = F_x.$$

As an application of this formulation, consider a particle falling freely near the surface of the earth. The approximation is made that the gravitational force is a constant, and all other forces are neglected. The rate of change of kinetic energy with distance is then a constant, namely mg. The change in kinetic energy when the object falls through a height, h, is thus mgh.

It is no longer necessary to assume that the body is falling in a straight line, nor is it necessary to restrict the discussion to motion in one dimension. The definition of kinetic energy, for the three-dimensional case, looks the same as it did before. In general, when Cartesian coordinates are used, the square of the speed is given by the equation

$$v^2 = v_x^2 + v_y^2 + v_z^2.$$

It is now shown that the change in kinetic energy for a body that falls through a vertical distance, h, is still given by the formula mgh, no matter what its horizontal motion is. This comes from the fact that when there is no force in the horizontal direction, the horizontal components of velocity remain unchanged. (It may be noted that the fact that the earth is a sphere is being neglected here.)

The formula for the motion just considered can be written as

$$\text{change in K} \equiv \Delta K = Wh = mgh$$

Note that the RHS of this equation involves the product of a force with a distance. This product is known as the work done on the body. In the English system, the unit of work is the foot-pound. The work done on a body yields its change in energy. It is only for the case of a constant force that work can be defined this way. In general, it is necessary to break the path of the particle into small sections and calculate the total work by adding up the contributions from each section.

The process described for calculating work is almost the opposite of the one used in defining a derivative. Recall the tabular method that was used to estimate the velocity and acceleration of a falling object, before calculus was applied. The tabulated position at different times was used to calculate the velocity. Then the tabulated values of the velocity were used to calculate the acceleration. If the acceleration had been known instead of the position the table could have been determined by working from bottom to top. It would have been necessary to know the velocity at one instant. The changes in the velocity could then be used to determine the remaining velocities. In the same way it would be necessary to know some position in order to use the changes to find the positions at other times. The generalization of these ideas to incorporate infinitesimal quantities is known as calculus. The generalization of working from top to bottom is called differential calculus. The generalization of working from bottom to top is called integral calculus. It should be clear that the definition of work involves an integral.

Potential Energy

The work done by the force of gravity near the surface of the earth is the product of the weight force with the distance an object falls. No matter how the object moves, the only thing that counts is the vertical distance through which it descends. It follows that if an object returns to its starting position, the work done by the gravitational force is zero, regardless of the path taken. Such a force is said to be conservative. Not all force fields are conservative. In some types of particle accelerators, a charged particle goes around and around in an orbit, picking up kinetic energy in every circuit. Such a force field is not conservative. For a conservative force field, it is not necessary to know the path taken by a particle to calculate the work done when it changes position from one point to another. Consider some arbitrary points in space called A and B. The work done on any conceivable path between the two points is the same as on any other. Imagine there were a difference in the work done by the force along one path between A and B and the work done along the other. In that case, the work done in going from A to B and back to A again could be different from 0. All that is necessary is to take one path from A to B and the other from B to A.

The concept of potential energy is a bookkeeping idea associated with conservative force fields. Take the case of the gravitational force field near the earth. When an object is situated on a mountaintop, for example, it has the possibility of gaining kinetic energy on coming down, because of the work done by the gravitational force. It has the potential to gain energy by descending. Instead of saying that a falling body gains energy as it falls, it can be said that it loses potential energy as it gains kinetic energy. The kinetic energy, the energy due to motion, is thus only considered as part of the total energy of the system. In order to assign a number to the potential energy, it is necessary to pick some reference point where it is assigned the number 0. For example, the potential energy of an object of mass m can be assigned the value of 0 at sea level. When it is at an elevation z above sea level, its value is mgz. A symbol often used for potential energy is V. Thus, it is useful to write

$$V = mgz$$

for the gravitational potential energy of an object of mass m near the surface of the earth. The total energy, E, of an object is the sum of the potential energy and the kinetic energy,

$$E = K + V.$$

For a freely falling body, the energy is conserved. This has very little to do with our present-day notion of energy conservation. In Newton's day, energy was not conserved, in general. Frictional forces slowed things down, and other types of forces were capable of speeding things up without loss of potential energy. Energy conservation was just a useful bookkeeping device. The numerical value of the energy has no significance, since it depends on where the potential energy is set to 0.

The force that an ideal spring exerts on an object is a conservative one. The potential energy of such a spring is given by the expression

$$V = \frac{1}{2}kx^2.$$

One can verify this by virtue of the following argument: The potential energy of a spring with a stretch x represents the work necessary to produce that stretch. As has been pointed out, the calculation of this work involves an integral, a process that can be viewed as the opposite of differentiation. In an analogous way, the potential energy involves an integral of force with respect to position so that force involves a derivative of the potential energy with respect to position. A negative sign is necessary in order to keep the record straight. In order to verify that the formula for the potential energy is correct, it is sufficient to differentiate the expression. The formula for the force is

$$F = -\frac{d(kx^2/2)}{dx} = -\frac{1}{2}k\frac{d(x^2)}{dx} = -\frac{1}{2}k(2x) = -kx.$$

This is Hooke's law for springs. Application of the same rule to the gravitational potential energy leads to

$$F = -\frac{d(mgz)}{dz} = -mg\frac{dz}{dz} = -mg.$$

The minus sign in the first equation indicates that the force is up when the position of the object is down. The minus sign in the second equation indicates that the force is downward, because z is positive in the upward direction. The formula for the potential energy of a spring can be used to obtain the speed of an object in SHM from a knowledge of x and the amplitude A. The kinetic energy is clearly 0 when x = A so that the total energy is just the potential energy when x = A. In other words:

$$E = \frac{1}{2}kA^2.$$

If friction is negligible, it is found that

$$\frac{1}{2}mv^2 + \frac{1}{2}kx^2 = \frac{1}{2}kA^2.$$

It is a simple matter to solve this equation for v if all the other variables are known.

It seems as though energy conservation can be used only when conservative forces are involved. This isn't quite true. Take the example of the pendulum consisting of a mass on a string. In addition to the gravitational force, there is the tension force from the string. Note that this force does no work since it is perpendicular to the motion. It is not responsible for any change in speed. It is not necessary to limit the motion of a pendulum to small oscillations in order to make use of the conservation of energy.

In the discussion of the ballistic pendulum in the previous chapter, it was stated that the speed of the bullet before impact could be determined if the speed of the pendulum could be found after impact. From the discussion of energy, it is clear that all that has to be measured is the maximum height gained by the suspended sandbox. Momentum conservation and energy conservation are both useful concepts. Newton's laws have interesting consequences.

Readers are encouraged to apply these ideas to various situations. For example, when an object slides or rolls on a surface, the contact force can be broken into two components. The perpendicular component is called the normal force (normal means perpendicular in mathematics). The tangential component is called the frictional force. Consider roller coasters, luges, and bobsleds and ask yourself, "Where is the speed the greatest? What is the maximum speed possible if its initial speed is known?" These ideas about energy can be applied to planetary motion. Note that the gravitational potential energy expression that has been used till now is only valid near the surface of the earth. It assumes that the gravitational force on an object doesn't depend on altitude.

The Scalar Product

The rate that energy is being delivered to a body is called the power input. In the MKS system, the unit of power is the watt, one joule per second. In the English system, it is the foot-pound per second. One horsepower is 550 foot-pounds per second. The power imparted to a particle by a force involves the product of the force with the body's velocity. This point has to be clarified since power is a scalar, and force and velocity are vectors. The product is designated by a dot:

$$P = \mathbf{F} \cdot \mathbf{v}.$$

It is known as a scalar product or dot product. In order to evaluate this product, the two vectors should be imagined as arrows that are drawn tail to tail, as in the figures below. The power is then given by

$$P = Fv \cos\theta.$$

It often helps to think of this expression in one of two different ways. One way is to think of it as the product of the magnitude of the force with the component of the velocity in the direction of the force. The first figure illustrates this. The velocity vector is expressed as the vector sum of its horizontal and vertical components. F is then multiplied by the parallel component.

The second way is shown in the second diagram, in which the force is resolved into two components. The speed is then multiplied by the parallel component of the force.

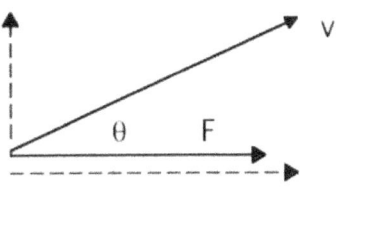
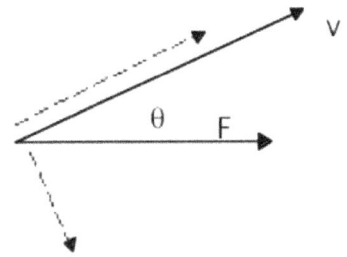

The first way is preferred when dealing with a constant force such as that due to gravity near the surface of the earth. In that case, the weight is multiplied by the downward component of the velocity. In this way, it is seen that the power being delivered by the gravitational force is the weight times the rate of decrease of altitude.

Scalar products occur often in physics. It is worth the effort to examine some of the properties. Consider an arbitrary sum of vectors—for example, the three vectors **A**, **B**, and **C**. It is important to show that the scalar product of the sum of these vectors with an arbitrary vector, **S**, satisfies the equation

$$(\mathbf{A} + \mathbf{B} + \mathbf{C}) \cdot \mathbf{S} = \mathbf{A} \cdot \mathbf{S} + \mathbf{B} \cdot \mathbf{S} + \mathbf{C} \cdot \mathbf{S}.$$

In order to see this, consult the following diagram. Imagine that the x-axis has been chosen to be parallel to the vector **S**. The easiest way to sum vectors pictorially is to join them head to tail, as shown in the figure. The vector **R** is, thus, the sum (or resultant) of the vectors **A**, **B**, and **C**. It is only necessary to show that the sum of the x-components of the individual vectors equals the x-component of the resultant. This is obvious from the diagram. This result can be extended to the situation where **S** consists of a sum of vectors. It should be clear that scalar products behave like ordinary products between numbers. It can be shown from the example that the power associated with the resultant of a bunch of forces is the sum of the powers from the individual forces.

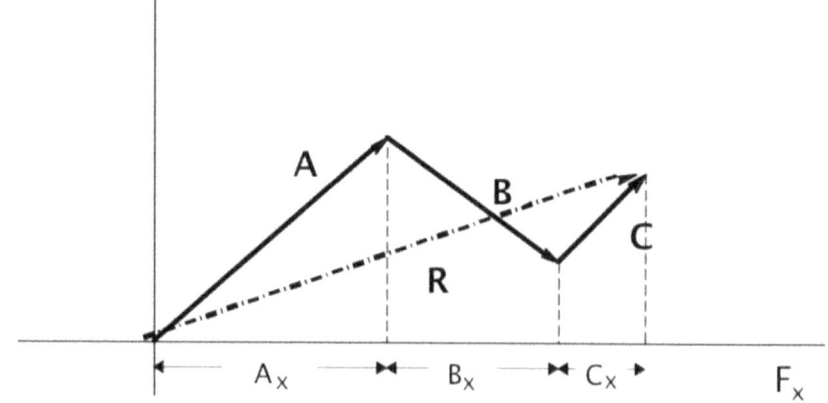

Although the illustration is two-dimensional, the arguments just used are valid in three dimensions. This result can be used to express the dot product of vectors in terms of their Cartesian components.

Consider a three-dimensional Cartesian coordinate system, x, y, and z. Construct three vectors, **i**, **j**, and **k**, which are parallel to the x-, y-, and z-axes, respectively. They each have unit length. They thus have the following properties:

$$\mathbf{i} \cdot \mathbf{j} = \mathbf{i} \cdot \mathbf{k} = \mathbf{j} \cdot \mathbf{k} = 0, \quad \mathbf{i} \cdot \mathbf{i} = \mathbf{j} \cdot \mathbf{j} = \mathbf{k} \cdot \mathbf{k} = 1.$$

The first set of equalities expresses the fact that these vectors are perpendicular to one another. The second set of equations expresses the fact that they have a length of one unit. They are dimensionless, so that if

$$\mathbf{F} = F_x\mathbf{i} + F_y\mathbf{j} + F_z\mathbf{k},$$

the components F_x, F_y, and F_z will have units (such as newtons, or pounds). In general, any vector can be expressed in terms of the unit vectors, **i**, **j**, and **k**. Note that

$$\mathbf{F} \cdot \mathbf{i} = \mathbf{i} \cdot \mathbf{F} = F_x.$$

The other components are related in a corresponding way.

Consider an arbitrary pair of vectors, **A** and **B**, and form their scalar product:

$$\mathbf{A} \cdot \mathbf{B} = \mathbf{A} \cdot (B_x\mathbf{i} + B_y\mathbf{j} + B_z\mathbf{k})$$
$$= B_x(\mathbf{A} \cdot \mathbf{i}) + B_y(\mathbf{A} \cdot \mathbf{j}) + B_z(\mathbf{A} \cdot \mathbf{k})$$
$$= A_xB_x + A_yB_y + A_zB_z.$$

This is a very useful result. Note that the scalar product of a vector with itself is the square of its magnitude. (This follows from the original definition of the scalar product and the definition of the cosine.) Thus, the magnitude, A, of the vector **A** is given by

$$A = \sqrt{A_x^2 + A_y^2 + A_z^2}.$$

Law of Cosines

A mathematical theorem that is a generalization of the Pythagorean theorem will now be derived. It is applicable to any triangle and will prove useful in the theory of elliptical planetary orbits.

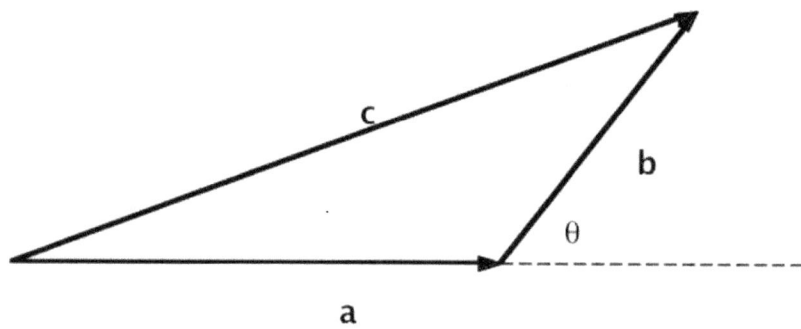

The triangle in the figure consists of sides of lengths a, b, and c. It should be clear that $\mathbf{a} + \mathbf{b} = \mathbf{c}$. If you take the scalar product of the vector $\mathbf{c}$ with itself, you find

$$(\mathbf{a}+\mathbf{b})\cdot(\mathbf{a}+\mathbf{b}) = \mathbf{c}\cdot\mathbf{c} = c^2.$$

The LHS of this equation can be written as

$$LHS = a^2 + b^2 + 2\mathbf{a}\cdot\mathbf{b}.$$

In order to evaluate the dot product, note that if the vectors a and b had been drawn tail to tail, theta would be the angle between them. It follows that

$$a^2 + b^2 + 2ab\cos\theta = c^2.$$

This is known as the law of cosines. It reduces to the Pythagorean theorem when theta is a right angle. A derivation of this law without the concept of scalar products is much more cumbersome.

Another mathematical equation that will prove useful is one that governs the cosine of the sum of two angles. This equation is also easily derived by making use of scalar products. Consider the diagram below,

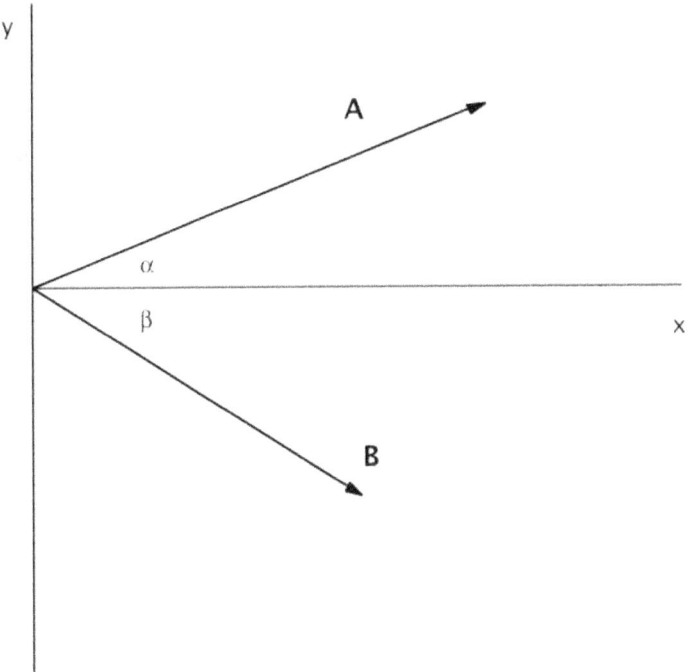

in which the Greek letters α and β are alpha and beta respectively. The scalar product of A and B can be expressed by two different formulae:

$$\mathbf{A} \cdot \mathbf{B} = AB\cos(\alpha + \beta) = A_x B_x + A_y B_y.$$

Note that B_y is negative. The result is

$$AB\cos(\alpha + \beta) = (A\cos\alpha)(B\cos\beta) + (A\sin\alpha)(-B\sin\beta).$$

Division by AB yields

$$\cos(\alpha + \beta) = \cos\alpha\cos\beta - \sin\alpha\sin\beta,$$

a result that will be needed later.

It is worth pointing out that Newton never made use of these vector techniques. Advantage is being taken of mathematical developments that came much later in history. In particular, a great debt is owed to the work of J. Willard Gibbs (1839–1903), who is called the father of vector analysis.

Gravitational Potential Energy

Newton was able to deal with the elliptical orbits of planets by a technique that is difficult to understand. For the present, it will be better to make use of energy conservation. For this purpose, an expression for the potential energy coming from the sun's gravitational field is needed. In order to derive this, use is made of the relationship between force and potential energy. In particular, it is found that

$$-\frac{dV}{dr} = -\frac{GMm}{r^2}.$$

In this equation, G is the universal gravitational constant, M is the mass of the sun, m is the mass of the planet, and r is the distance between their centers. The negative sign on the RHS expresses the fact that the force is inward while r gets larger when the planet moves outward. For the sake of those readers who don't like derivations, the formula is

$$V = -\frac{GMm}{r}.$$

The proof of the above formula involves the rule for differentiating a product of two functions. This rule will be needed again in the next chapter. Instead of dealing with the variable r, the notation of chapter one is used. Let f be some function of time, f(t), which means that the numerical value of f is specified for all values of t. It is well behaved and has derivatives everywhere. Let g(t) also be such a function. The derivative of the product of these functions is defined by

$$\frac{d(fg)}{dt} = \text{the limit as s approaches zero of } \frac{f(t+s)g(t+s) - f(t)g(t)}{s}.$$

The RHS of this equation can be rewritten as

$$\frac{f(t+s)g(t+s)-f(t+s)g(t)+f(t+s)g(t)-f(t)g(t)}{s}.$$

Note that two terms that sum to 0 have been inserted into the numerator. This expression can be written as

$$f(t+s)\frac{g(t+s)-g(t)}{s}+\frac{f(t+s)-f(t)}{s}g(t).$$

Examine what happens as s approaches 0. The first term approaches f(t), and the fractional terms yield time derivatives. It follows that the derivative of an arbitrary product takes the form

$$\frac{d(fg)}{dt}=f\frac{dg}{dt}+\frac{df}{dt}g.$$

It is worthwhile to remember this result, as it comes up frequently. This general rule is applied to the situation in which f = 1/t, and g = t, so that the product fg = 1, namely, a constant. Since the time derivative of a constant is 0, it follows that

$$0=-\frac{1}{t}\frac{dt}{dt}+\frac{d(1/t)}{dt}t=1/t+t\frac{d(1/t)}{dt}.$$

It is found that

$$\frac{d(1/t)}{dt}=-\frac{1}{t^2}.$$

The variable t can be replaced by r so that it can be seen that the given potential energy function -GMm/r leads to the correct force law.

In order to get some insight into the meaning of this potential energy function, it is applied to the gravitational field of the earth. According to the general formula for the gravitational force, the weight of an object of mass m on the earth's surface is

$$W=mg=GMm/R^2,$$

where R is the radius of the earth and M is its mass. The potential energy of the object at this point is

$$V = -GMm / R = -WR.$$

This equation is equivalent to the following statement: The energy needed for an object to escape from the earth's gravitational field is equivalent to the energy needed to lift it about 4,000 miles (the radius of the earth) if its weight was constant. This calculation is based on the assumption that the object has no kinetic energy on the earth's surface. Since the earth is rotating with respect to an inertial frame this is equivalent to assuming the object sits at one of the poles.

Chapter 5
The Planets

The material covered so far is enough to illustrate the genius of Newton. Simple-sounding laws had far-reaching consequences. The ideas led to mathematical techniques that proved to be extremely powerful in their own domain of mathematics. However, the work that really brought Newton to the attention of the astronomers of his day was his account of planetary motion. Johannes Kepler had summarized the observations of the Danish astronomer Tycho Brahe in the form of three laws. They are as follows:

1. Planetary orbits are ellipses with the sun located at one focus.

2. As a planet moves along its orbit, the line from the sun to the planet sweeps out equal areas in equal times.

3. The square of the period of any orbit is proportional to the cube of the length of the semi-major axis. (Recall that the period is the time it takes to complete an orbit.)

Kepler's Second Law

Of the three laws, the second is the easiest to understand. It is a consequence of the gravitational force being directed toward the center of the sun. Imagine a line that extends from the center of the sun to the center of some planet. As the planet moves, this line can be thought of as acting like the windshield wiper of a car. It sweeps out an area of space.

The length of this wiper blade is r. If the tip of it went around in a complete circle, the wiper would sweep an area of πr^2. A complete circle corresponds to an angle of 2π radians. Thus, when it turns through an angle θ, the wiper sweeps out an area $r^2\theta/2$. If the rate at which the angle θ is changing is ω, the rate at which the area is being swept is $r^2\omega/2$. This formula is correct even if r depends on time. Kepler's second law is equivalent to the following equation:

$$\frac{d(r^2\omega)}{dt} = 0.$$

This equation implies that the term in parentheses is constant.

The proof of this equation involves the general expression for the acceleration in polar coordinates. Since the force on a planet is directed toward the sun, its acceleration must be in that direction also. The perpendicular component of the acceleration must be 0. In order to get an expression for the acceleration, it is useful to remember the following properties of any vector quantity. It can change because its magnitude is changing. It can also change because its direction is changing. The first change is a vector pointing in the same direction as the vector itself. The second change is perpendicular to the vector. This situation was encountered in dealing with uniform circular motion.

This last idea is applied to the rate of change of the position vector in order to calculate the velocity. The component of the velocity vector in the radial direction is due to the changing rate of r, namely dr/dt. The tangential component is the one encountered in uniform circular motion, namely ωr. (It is called tangential because it is in the direction of the tangent to the circular orbit that you get when r is constant.) In summary, the radial and tangential components of the velocity vector are

$$v_r = \frac{dr}{dt}, \quad v_t = \omega r.$$

The calculation of the acceleration is more complicated. Both components of the velocity vector may be changing. The changing magnitude of the radial component is in the radial direction. The changing direction of the tangential component is inward, as was seen in uniform circular motion. The net radial acceleration is given by the formula

$$a_r = \frac{dv_r}{dt} - \omega^2 r.$$

This could be used in Newton's second law because the force of the sun on the planet is in the radial direction, although it is negative (inward).

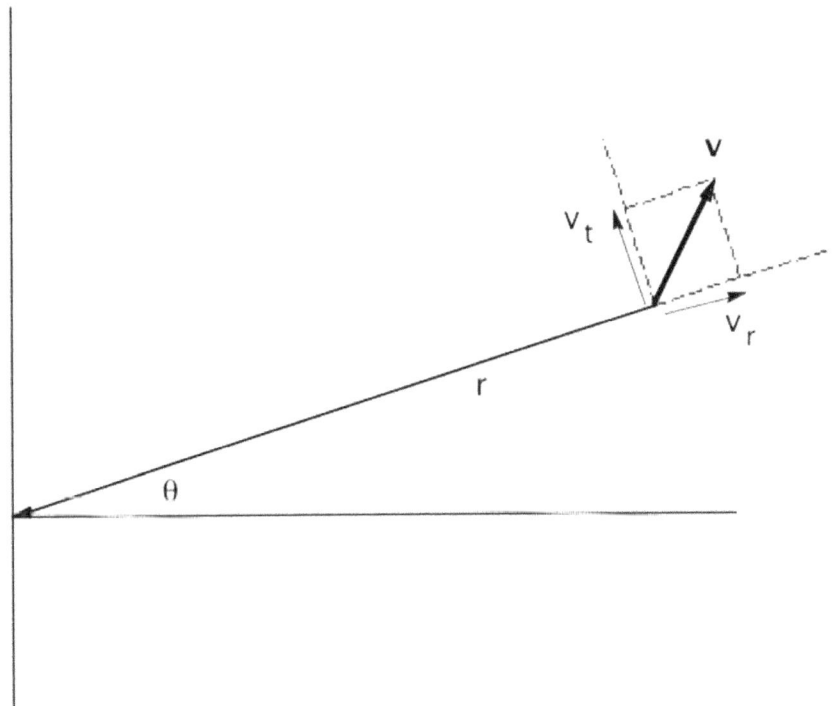

The tangential part of the acceleration has two contributions also. One contribution comes from the changing direction of the radial component of velocity. The other comes from the changing magnitude of the tangential component of velocity. The end result is

$$a_t = \omega \frac{dr}{dt} + \frac{d(\omega r)}{dt}.$$

Note that ω is not necessarily a constant in planetary motion. It can depend on time. Note also that this component of acceleration is 0 since there is no component of force in this direction. Use of the rule for differentiating a product leads to

$$2\omega \frac{dr}{dt} + r \frac{d\omega}{dt} = a_t = 0.$$

It will now be shown that the LHS can be simplified to

$$\frac{1}{r} \frac{d(r^2 \omega)}{dt}.$$

In order to make the comparison, it is again necessary to differentiate a product,

$$\frac{1}{r} \frac{d(r^2\omega)}{dt} = \frac{1}{r}(2r \frac{dr}{dt} \omega + r^2 \frac{d\omega}{dt}) = 2\omega \frac{dr}{dt} + r \frac{d\omega}{dt}.$$

Since the quantity on the RHS has been shown to be 0, the conclusion is that $r^2\omega$ is a constant. This accounts for Kepler's second law.

Actually, the quantity $r^2\omega$ is of great interest in mechanics. It is related to the angular momentum of the planet about the sun. The actual formula for the angular momentum is

$$L = mr^2\omega.$$

Angular momentum plays a role similar to that of momentum for a rotating system. It can be said that Kepler's second law is a consequence of the conservation of angular momentum. Use will be made of this and the conservation of energy in verifying Kepler's other laws.

WARNING!

The material in the next two sections is difficult. It is included for completeness, to show the power of Newton's laws. You may skim through these sections. They are not necessary for understanding the material that comes later. If you like challenges, it is recommended that you use pencil and paper, copying down formulas that are needed from earlier sections. For example, use is made of a formula that was proven when dealing with SHM, namely,

$$\frac{d(\cos\theta)}{d\theta} = -\sin\theta.$$

Ellipses

Whereas a circle has one special point, known as the center, for an ellipse there are two points, known as the foci. Consider the figure below. The two foci are shown lying on the x-axis. An ellipse is defined as the locus of all points in a plane, such that the sum of their distances from the foci is a constant. This constant is 2a. From the figure, it is seen that r + ρ = 2a (ρ is the Greek letter *rho*, pronounced *roe*). If this ellipse is to correspond to a planetary orbit, you can imagine the sun to be at the focus on the right. In this case, the letter r will correspond to the planet's distance from the sun. The ellipse is symmetrical about the x-axis, as well as the y-axis, in the diagram below. (It is assumed here that the origin of Cartesian coordinates is halfway between the foci.) The major axis is the horizontal axis of symmetry. Its length is 2a. Since the origin of the Cartesian coordinates is not at the sun, the usual relationship between the polar and Cartesian coordinates does not hold.

The shape of an elliptical orbit is completely determined by the distance between the foci and the parameter, a. The distance between the foci is usually expressed as 2ea, where the letter *e* is called the eccentricity of the orbit. If e = 0, the two foci coincide so that r and ρ are the same. In this case, the ellipse becomes a circle of radius a. A circle is a special case of an ellipse. The eccentricity is some number between 0 and 1.

The point of closest approach to the sun is called perihelion. In terms of the symbols already used, this distance is (1- e)a. The farthest distance is called aphelion and is (1 + e)a. The perpendicular symmetry axis is the minor axis. The length of the semi-minor axis is called b. The value of b can be found from the following analysis. The lowest point is the same distance from each focus. This distance is therefore a. It is the hypotenuse of a right triangle with one side of length b and another of length ea. Use of the Pythagorean theorem leads to the relationship

$$b^2 = (1-e^2)a^2.$$

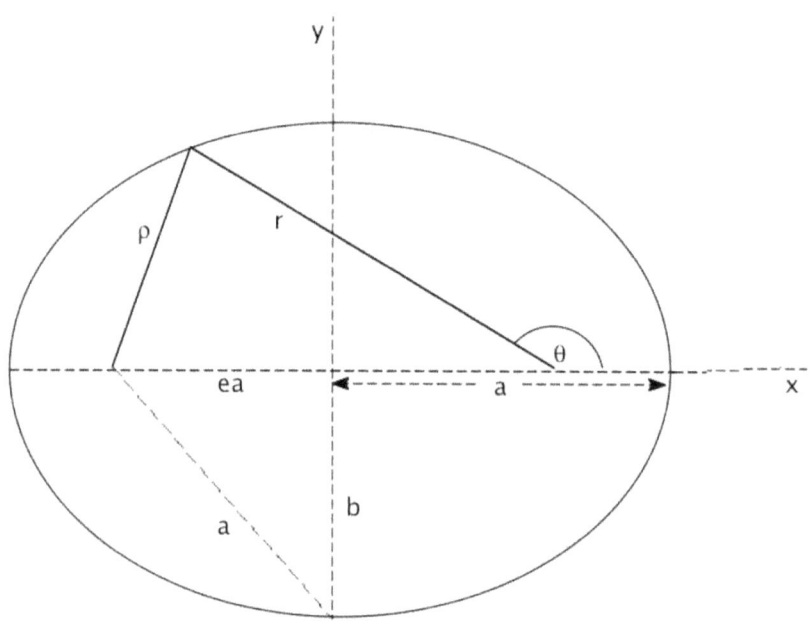

An equation for this ellipse in terms of the polar coordinates r and θ may be found by using the law of cosines. The law is applied to the triangle of sides r, ρ, and 2ea. Use is also made of the fact that ρ = 2a - r, since r + ρ = 2a. This leads to the following:

$$(2a-r)^2 = (2ea)^2 + r^2 + 2r(2ea)\cos\theta.$$

This is equivalent to

$$4a^2 - 4ar + r^2 = 4e^2a^2 + r^2 + 4ear\cos\theta.$$

Subtract r² from each side. Then divide through by 4a². This leads to

$$1 - \frac{r}{a} = e^2 + \frac{r}{a}e\cos\theta.$$

This can be simplified further:

$$1 - e^2 = r(1 + e\cos\theta)/a.$$

Solving this equation for r yields

$$r = \frac{a(1-e^2)}{1+e\cos\theta}.$$

However, it will be more convenient to have the angular dependence in the numerator. For this reason it is better to write the equation in the following form:

$$\frac{1}{r} = \frac{1+e\cos\theta}{a(1-e^2)}.$$

You may check that this equation yields the correct values for the perihelion and aphelion of the orbit. In order to see this, it is helpful to recall the following:

$$1 - e^2 = (1 - e)(1 + e), \cos 0 = 1, \text{ and } \cos \pi = -1.$$

Theory of Planetary Orbits

It is useful to outline the plan of attack, as it is a bit tedious. It will be assumed that only the angle, θ, depends explicitly on the time, t. The time dependence of r comes about indirectly, because r depends on θ, according to the orbital equation. The method makes use of the conservation of energy and of angular momentum. The objective is to get an equation that involves only θ and some constants. It will be seen that the only way it will be possible to solve this equation for an arbitrary angle will involve making the proper choice of these constants.

The velocity vector of a particle in polar coordinates has two perpendicular components. As shown earlier, these are ωr and dr/dt. The square of the planet's speed is just the sum of the squares of these two components. Since the gravitational force of the sun is conservative, the total energy is a constant. This yields the following,

$$\frac{m}{2}[(\omega r)^2 + (\frac{dr}{dt})^2] - \frac{GMm}{r} = E,$$

where M is the mass of the sun. Recall that r depends on t indirectly. Thus:

$$\frac{dr}{dt} = \frac{dr}{d\theta}\frac{d\theta}{dt} = \omega\frac{dr}{d\theta}.$$

Use of this result in the previous equation leads to

$$\frac{m\omega^2}{2}[r^2 + (\frac{dr}{d\theta})^2] - \frac{GMm}{r} = E.$$

The angular velocity varies with position for this motion. However, it has been shown that the angular momentum, L, is a constant. It is thus possible to get an equation in which the only variables involve the orbital position:

$$L = mr^2\omega, \qquad \text{so that } \omega^2 = \frac{L^2}{m^2 r^4}$$

The result is

$$* \quad \frac{L^2}{2m}[\frac{1}{r^2} + \frac{1}{r^4}(\frac{dr}{d\theta})^2] - \frac{GMm}{r} = E . *$$

This equation is starred for reference purposes. All the time dependence has been eliminated. The starred equation involves some constants and the position variables. It is necessary to show that an elliptical orbit satisfies this equation, provided a and e are appropriately related to E and L. In other words, a solution is assumed in the following form:

$$\frac{1}{r} = \frac{1+e\cos\theta}{a(1-e^2)}.$$

The problem reduces to showing that this equation is consistent with the starred equation for an appropriate value of E and of L. The process is a bit tricky. The first task is to eliminate the coordinate r from the starred equation by making use of the orbital equation. Since one of the terms depends on dr/dθ, it is useful to differentiate both sides of the orbital equation with respect to θ. For this purpose, recall the following relationship:

$$\frac{d(\cos\theta)}{d\theta} = -\sin\theta.$$

This equation was first encountered in the section on SHM. Differentiation of the LHS of the orbital equation yields

$$\frac{d(1/r)}{d\theta} = \frac{d(1/r)}{dr}\frac{dr}{d\theta} = -\frac{1}{r^2}\frac{dr}{d\theta}.$$

Equating this to the derivative of the RHS of the orbital equation leads to

$$-\frac{1}{r^2}\frac{dr}{d\theta} = \frac{e\sin\theta}{a(1-e^2)}.$$

Now look at the starred equation. The first term inside the brackets is the square of 1/r. The second term inside the brackets is the square of the term just considered. Both of these terms have been expressed as functions of the angle variable and some constants. The sum of these terms can be simplified as follows:

$$\frac{1}{r^2} + \frac{1}{r^4}\left(\frac{dr}{d\theta}\right)^2 = \frac{1+2e\cos\theta+e^2(\cos^2\theta+\sin^2\theta)}{a^2(1-e^2)^2}$$

$$= \frac{1+2e\cos\theta+e^2}{a^2(1-e^2)^2}.$$

This expression is inserted into the starred equation. In addition, the potential energy term is expressed in terms of theta. The resulting equation now involves only one variable, as shown:

$$\frac{L^2(1+2e\cos\theta+e^2)}{2ma^2(1-e^2)^2} - GMm\frac{1+e\cos\theta}{a(1-e^2)} = E.$$

This equation can be brought into the following form:
$$A + B\cos\theta = E,$$
where A , B, and E are constants. Note that this is not an equation to determine the angle theta. It is an equation that must be satisfied for all the angles in the orbit. The only way this equation can be valid for all angles is for the constant B to be 0. Setting the coefficient of cosθ to 0 yields

$$\frac{L^2 e}{a^2 m(1-e^2)^2} - \frac{GMme}{a(1-e^2)} = 0.$$

This relates the angular momentum to certain features of the orbit, namely,

$$L^2 = GMm^2 a(1-e^2).$$

This value can be put back into the earlier equation. The terms involving the angle, θ, will be absent. What remains is

$$\frac{GMm(1+e^2 - 2)}{2a(1-e^2)} = -\frac{GMm}{2a} = E.$$

If you were able to plow through all these equations, you should be congratulated. Newton used a different approach that wasn't any easier. The labor is justified, however. The importance of this solution cannot be exaggerated. His laws had been successful in taking the mystery out of the motion of heavenly bodies.

He had found a simple set of equations that other scientists could now apply to all sorts of situations. Newton's work didn't only have an impact on mathematicians and scientists. It affected philosophy and religion as well. No longer were angels needed to carry planets around in their strange paths. Their motion was no more mysterious than that of a rock thrown by some youngster. The astronomer Edmond Halley (1656–1742) was tremendously impressed by Newton's results. When he discovered that Newton could account for the motion of the planets, he urged Newton to publish *The Principia*, a book that contained all of Newton's work. Halley paid the publication costs. Newton stands out as a superstar among scientists. Only Albert Einstein and Charles Darwin have achieved comparable eminence.

Kepler's third law still has not been accounted for. This is the law relating the period, T, to the length of the major axis. Recall that L/2m is the rate at which the area is being swept out by the vector r. It follows that if A is the area of the elliptical orbit, the period is given by

$$T = 2mA / L.$$

Thus, a formula is needed for the area of an ellipse. If you are not interested in mathematical proofs, you might be happy to know that the formula is A = πab. (A proof is given in the next section.) It has been shown that b² = a²(1-e²) so that the period can be written as

$$T = 2\pi ma^2 \frac{\sqrt{1-e^2}}{L}.$$

You should square this equation and make use of the fact that L² = Gm²Ma(1-e²). You will find that

$$T^2 = \frac{4\pi^2 a^3}{GM}.$$

This is Kepler's third law! Not only does it state that the square of the period of planetary motion is proportional to the cube of the semi-major axis, it specifies the constant of proportionality. When Henry Cavendish measured G in 1798, he was able to determine the mass of the sun! His experiment is described on the Internet.

The Area of an Ellipse

Polar coordinates have been used for the description of ellipses. The use of Cartesian coordinates will provide new insights into some of their properties. The origin of the Cartesian system is chosen to be at the center of the ellipse. It is different from that of the polar coordinates system so that

$$x = r\cos\theta + ea, \quad y = r\sin\theta.$$

Note that
$$(x-ea)^2 + y^2 = r^2(\cos^2\theta + \sin^2\theta) = r^2.$$
The equation for an ellipse in polar coordinates was shown to be
$$\frac{1}{r} = \frac{1+e\cos\theta}{a(1-e^2)}.$$
Multiplication of each side by $a(1-e^2)r$ yields
$$a(1-e^2) = r + e(x-ea)$$
$$a(1-e^2) - e(x-ea) = r.$$
This equation simplifies to
$$a - ex = r.$$
Make use of the equation for r²:
$$r^2 = (x-\varepsilon a)^2 + y^2 = (a-\varepsilon x)^2.$$
This equation simplifies to
$$x^2(1-e^2) + y^2 = a^2(1-e^2).$$
Recall that $(1-e^2)a^2 = b^2$. Division by this term leads to
$$\frac{x^2}{a^2} + \frac{y^2}{b^2} = 1,$$
as the equation for an ellipse in Cartesian coordinates.

From this equation, scaling arguments can be used to show that the area of the ellipse under consideration is given by the formula
$$A = \pi ab.$$
In the present context, scaling implies a type of stretch in some direction. When a photograph is enlarged, it is scaled uniformly in all directions. When widescreen TV is used, an image is expanded in only one direction. Assume that b is smaller than a, as is the case when 2a is the length of the major axis. Imagine stretching the figure in the y direction. The coordinate y is replaced by y' (read this as y prime) according to the rule
$$y' = (a/b)y.$$
For example, if a = 2b, the diagram is expanded uniformly about the x-axis by a factor of 2 in the vertical direction. The equation for this new figure is

$$\frac{x^2}{a^2} + \frac{y'^2}{a^2} = 1, \quad \text{or equivalently } x^2 + y'^2 = a^2.$$

It should be clear that this is the equation of a circle of radius a, whose area is πa^2. Ellipses are related to circles in that they can be transformed into one another by a stretch or a compression in some direction. In the case just considered, the stretch was a/b so that the area of the ellipse is given by πab, as was assumed earlier.

Actually, the claim that elliptical orbits are the only ones possible in the sun's gravitational force field is not strictly correct. It has already been noted that the energy is negative for planetary orbits. Other orbits should be possible if the energy is positive. Note that the potential energy is 0 at an infinite distance from the sun. Thus, negative energies are associated with objects that cannot reach infinity. That should be clear from the fact that the kinetic energy cannot ever be negative. However, if a body had a total positive energy, the kinetic energy would still be positive as it got sufficiently far away as to be essentially beyond the reach of the sun. The positive energy solutions would thus apply to bodies that have been launched from a planet or some other source with enough energy to escape the solar system. They would also apply to a body that might have come into our solar system from far away. It turns out that Newton's laws are capable of dealing with these trajectories as well as those of planets, meteors, and comets. These positive energy solutions turn out to be hyperbolas. The equation for such a trajectory resembles the ones for ellipses. It is

$$\frac{1}{r} = \frac{1 - e\cos\theta}{e^2 - 1}.$$

This equation looks very much like the one for an ellipse. One difference is that the parameter e is larger than one in this case. The numerator becomes 0 when

$$\cos\theta = 1/e.$$

When this equation is satisfied, the body is at an infinite distance from the sun. There is no solution for angles smaller than that because r cannot be negative.

Before leaving the topic of orbits, it should be realized that el-liptical orbits, which include circular ones, are the only kind possi-ble for low-energy solutions to the inverse square force law. If someone asks you, "What kind of a trajectory do you obtain when a rock is thrown, neglecting air resistance?" you should answer, "It is a portion of an ellipse." You are implying that it is being viewed from an inertial frame of reference, not the rotating earth. Be care-ful, however. If the question is part of a physics exam, you are safer saying it is a parabola. This is the answer that is considered correct in most elementary physics classes. That answer would be correct if the earth did not rotate, if it was flat, and if g didn't vary with alti-tude.

The treatment of planetary motion dealt with the motion of the center of mass of a planet. The techniques, which have already been discussed, need to be extended in order to deal with the behavior of spinning objects. Some rotational phenomena, like those associated with gyroscopes, can be quite complex. For example, if you have ever tried to change the orientation of the axis of a rapidly spinning wheel, you might have noticed that it usually doesn't shift in the expected direction. The behavior of a body that is spinning about an axis that is fixed in space is much simpler. Even these rotations can be complicated if the spinning object isn't symmetrical. It is easiest to start off the study with the behavior of symmetrical bodies such as wheels.

Actually, Newton's three laws of motion are not quite sufficient to deal with the rotational motion of solids. In order to see this, imagine you had a phonograph turntable on which were glued some special substance at opposite points on the rim. Suppose these objects could exert forces on one another, as shown in the figure below

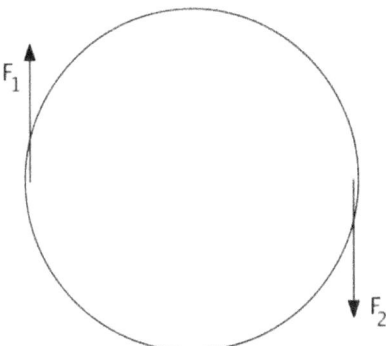

Could these forces be an action-reaction pair? They satisfy Newton's third law by being equal in magnitude and in opposite directions. Nevertheless, physicists believe this situation to be impossible. If it were possible, the turntable would start to turn, and in the absence of friction, it would turn faster and faster. The kinetic energy and the angular momentum of all the particles in this system would increase. This is generally considered to be impossible. Perhaps Newton didn't think it was necessary to mention the possibility of such a strange force. There were no known action-at-a-distance forces that behaved anything like the one shown in the diagram. A statement to the effect that this situation is impossible is contained in a law not directly attributed to Newton, called the conservation of angular momentum.

Conservation of Angular Momentum

Many mathematical concepts that physicists use today were not known in Newton's day. In particular, most of our understanding of vectors was due to developments in the nineteenth century. The book *Vector Analysis*, by the brilliant American physicist Willard Gibbs (1839–1903), had a great impact on physicists and mathematicians. Vectors will be used to deal with quantities that once were handled differently. It makes the analysis much simpler.

The Cross Product

There are two common ways of combining vectors, other than addition. You have already encountered the scalar (or dot) product. The result of combining two vectors in that case resulted in a scalar. The second way of combining two vectors is known as the vector product, also called the cross product. As the name suggests, the result of a vector product is another vector. In order to define the cross product between vectors **A** and **B**, they are drawn tail to tail, just as was done for scalar products. Let the angle between these vectors be called θ. See the diagram below.

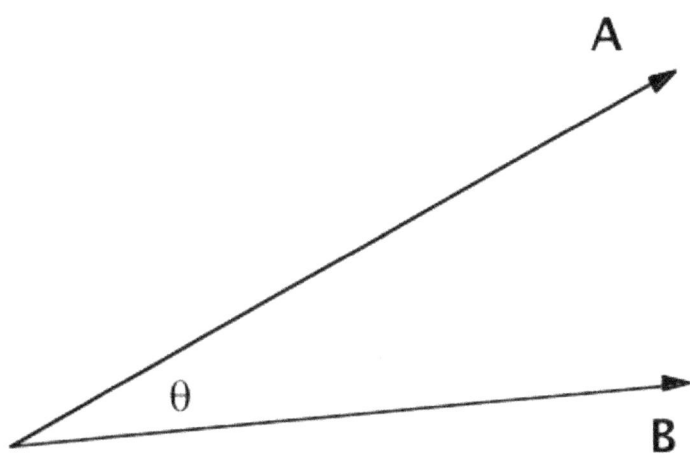

The easiest part of specifying a cross product is contained in the equation for the magnitude, C, of the cross product, **A** x **B** = **C**, given by

$$C = AB\sin\theta.$$

This differs from the scalar product in that the cosine is replaced by the sine of the angle. This number can be interpreted as arising from the product of the perpendicular component of vector **A** with the magnitude of **B**. Alternatively, it can be regarded as the product of the perpendicular component of **B** with the magnitude of **A**. The direction of **C** is defined to be perpendicular to the plane containing **A** and **B**. In other words, it is a vector perpendicular to both **A** and **B**. That still is not enough, since it can be directed into the page, or in the opposite direction. The correct answer depends on the order in which **A** and **B** is written. In the present case, **A** came first (to the left of **B**). In that case, the vector **C** is directed into the page. In order to reach this conclusion, it helps to be familiar with right-handed threads on screws or lightbulbs. If you were to rotate the first vector toward the second (the shortest way possible), you would be rotating **A** clockwise. A right-handed thread would advance into the page. The cross product denoted by **B** x **A** would be of the same magnitude but directed toward the reader. In general:

$$\mathbf{A} \times \mathbf{B} = -\mathbf{B} \times \mathbf{A}.$$

The properties of the cross product are similar to those of the scalar product in that

$$\mathbf{A} \times (\mathbf{C} + \mathbf{D}) = \mathbf{A} \times \mathbf{C} + \mathbf{A} \times \mathbf{D}.$$

$$\frac{d(\mathbf{A} \times \mathbf{B})}{dt} = \mathbf{A} \times \frac{d\mathbf{B}}{dt} + \frac{d\mathbf{A}}{dt} \times \mathbf{B}.$$

The first of these equations can be used to express the cross product of two vectors in terms of their Cartesian components. A right-handed coordinate system is assumed such that the unit vectors satisfy the relation $\mathbf{i} \times \mathbf{j} = \mathbf{k}$. It is true then that $\mathbf{j} \times \mathbf{k} = \mathbf{i}$ and $\mathbf{k} \times \mathbf{i} = \mathbf{j}$. Note that the cross product of two parallel vectors is 0. You should be able to verify that

$$\mathbf{A} \times \mathbf{B} = (A_y B_z - B_y A_z)\mathbf{i} + (A_z B_x - B_z A_x)\mathbf{j} + (A_x B_y - B_x A_y)\mathbf{k}.$$

The cross product is a useful tool when dealing with rotations. It can be used to calculate the velocity of a particle on a rotating object. The angular velocity is represented by a vector, ω, that points along the axis of rotation. It points in the direction that a right-hand screw would advance. Consider the diagram below. The letter O designates an arbitrary origin that is located on the axis of rotation of a rigid body. The vector **r** represents the position of a particle on the body with respect to this origin. At the instant shown, this particle is going through the circular motion of radius d and is moving into the page. Its speed is given by the product ωd. It is assumed that the angular velocity is given in radians per second. Note that d is given by r sin θ so that

$$v = \omega r \sin \theta.$$

It can be seen that

$$\mathbf{v} = \omega \times \mathbf{r}.$$

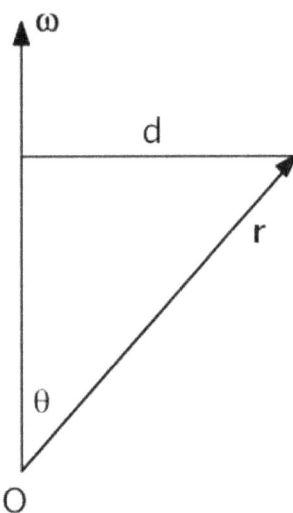

The same idea can be used whenever any vector is changing by virtue of a rotation. In the example just considered, the velocity represents the rate of change of position with time. If some vector, say **A**, is rotating with angular velocity ω and is not changing in magnitude, then the following equation holds:

$$\frac{d\mathbf{A}}{dt} = \omega \times \mathbf{A}.$$

Torque and Angular Momentum

When discussing Newton's second law, it is assumed that the law is being applied with respect to an inertial frame. It has already been noted that there are an infinite number of inertial frames. Quantities such as momentum and kinetic energy will then depend on what frame is chosen. The quantities that will be dealt with in this section will also depend on the choice of frame. Moreover, they will also depend on where the origin of coordinates is placed in such a frame. A vector quantity, known as torque, τ (Greek letter *tau*), can be defined with respect to some chosen origin. If a force,

F, is applied to a body at position **r**, the torque due to that force about the origin is

$$\tau = \mathbf{r} \times \mathbf{F},$$

where the origin is chosen is a matter of convenience. Sometimes the word *torque* is replaced by the phrase *the moment of the force.*

The angular momentum, **L**, of a particle with respect to an origin is defined by the equation

$$\mathbf{L} = \mathbf{r} \times m\mathbf{v}.$$

It is a generalization of the angular momentum associated with that of planetary motion as can be seen from the following: The angular momentum, L, about the sun, that was used for orbital motion was given by the formula

$$L = mr^2 \omega.$$

It was treated as a scalar, but it is preferable to think of it as the magnitude of the vector **L**. Note that the vector **L** is perpendicular to the plane of the orbit. The magnitude of this vector is the product of r with the component of momentum perpendicular to **r**. This latter quantity is mωr. Thus, the magnitude of **L** agrees with the earlier definition of angular momentum

The rate of change of the angular momentum of a particle about some origin equals the torque about that origin, as is shown by the following:

$$\frac{d\mathbf{L}}{dt} = \frac{d(\mathbf{r} \times m\mathbf{v})}{dt} = \frac{d\mathbf{r}}{dt} \times m\mathbf{v} + \mathbf{r} \times \frac{d(m\mathbf{v})}{dt}$$

$$= \mathbf{v} \times m\mathbf{v} + \mathbf{r} \times \mathbf{F} = \mathbf{r} \times \mathbf{F} = \tau.$$

The relationship between torque and angular momentum is analogous to the relationship between force and linear momentum.

The chapter began with an assumption of the impossibility of certain types of forces. Internal forces associated with a system of particles cancel one another, insofar as the torque about some origin is concerned. This leads to the law of the conservation of angular momentum, which can be stated as follows:

The angular momentum of a system of particles about some origin in an inertial frame is conserved in the absence of external torques on the system.

Alternatively:

The rate of change of the angular momentum of a system of particles about some origin in an inertial frame equals the resultant torque of all the <u>external forces</u> acting on that system.

The Role of the Center of Mass

The question sometimes arises as to what choice of origin is best for calculating the angular momentum or the torque on a system of particles. Two possible origins, O and O', are considered in the following diagram.

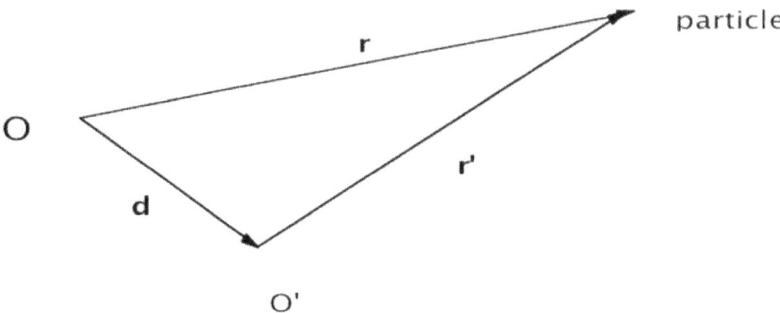

The position vector of a particle, as measured from the primed origin, is primed, while the other is not. From the diagram it is seen that

$$\mathbf{r} = \mathbf{d} + \mathbf{r}'.$$

Consider now a system consisting of an arbitrary number of particles. Only three particles are chosen for this example. Their angular momentum is first calculated with respect to O. Call this $\mathbf{L}_0$. It is found that

$$\mathbf{L}_0 = \mathbf{r}_1 \times m_1 \mathbf{v}_1 + \mathbf{r}_2 \times m_2 \mathbf{v}_2 + \mathbf{r}_3 \times m_3 \mathbf{v}_3.$$

Each position vector can be expressed in terms of the primed position vector and the constant vector $\mathbf{d}$. In this way, $\mathbf{L}_0$ can be expressed in terms of the angular momentum about O' plus a correction term. This correction term is

$$correction = \mathbf{d} \times (m_1\mathbf{v}_1 + m_2\mathbf{v}_2 + m_3\mathbf{v}_3).$$

It should be noted that the term in parentheses is the total momentum of the system. It is given by the total mass times the velocity of the center of mass. If the c.m. (center of mass) is stationary in this inertial frame, there is no correction. In this case, it is possible to pick a convenient origin and be sure that the result will be valid for any other fixed origin in this frame of reference. For example, the angular momentum of a car's tire about any point in the car is the same as it is about the center of the tire.

A claim similar to the one used for angular momentum can be made for the torque applied to a system of particles. The only thing that is different is that the momentum of the particle, m**v**, is replaced by the net force on it. In that case, the correction term involves the cross product of **d** with the resultant force on the system. If the c.m. is not accelerating, this term is 0 also. This simplifies the analysis of many situations.

Suppose that O' is not stationary in the given reference frame. It may even be accelerating. The formula for the angular momentum about the fixed origin, O, in terms of the angular momentum about O', is still valid. In particular, if O' is chosen at the c.m., the correction term has an easy interpretation. The correction is the same as the angular momentum of a particle with the mass of the body located at the c.m. For example, the angular momentum of the earth about the sun is the same as that of a particle with the earth's mass plus the angular momentum about the earth's c.m.

It is necessary to be careful in interpreting this result because the velocities that were used in the derivation were with respect to the inertial frame, not with respect to a frame moving with the c.m. However, it can be seen that this makes no difference. Primes are used to designate velocities with respect to the c.m., and **V** denotes the velocity of the c.m. Since the velocity is the derivative of the position with respect to time, and the derivative of a sum is the sum of the derivatives, it can be seen that

$$\mathbf{v}_i = \mathbf{V} + \mathbf{v}'_i, \quad i = 1 \text{ or } 2 \text{ or} \ldots..$$

The angular momentum about O' can now be written as a sum of terms, each of which has the following form:

$$m_i \mathbf{r}'_i \times \mathbf{v}_i = m_i \mathbf{r}'_i \times (\mathbf{V} + \mathbf{v}'_i)$$
$$= m_i \mathbf{r}'_i \times \mathbf{V} + m_i \mathbf{r}'_i \times \mathbf{v}'_i.$$

It will now be shown that when the total angular momentum is calculated, the primed velocities can be used instead of the unprimed ones. The term involving the velocity of the c.m. sums to 0. In order to see this, recall how the c.m. was defined. When summed, the first term on the RHS of the last equation involves the product of the total mass with the primed position vector of the center of mass. However, the c.m. is at the origin of the primed frame of reference. This term is 0.

The information contained in the last paragraph is useful in calculating the angular momentum of a rolling ball or cylinder. To obtain its angular momentum with respect to a point in a stationary frame of reference, the two contributions are added to one another. One comes from the motion of the c.m.; the other comes from the rotation about the c.m. It doesn't matter if the c.m. is accelerating.

It will be proved that the kinetic energy of any system of particles can be regarded as the sum of two parts: 1. the part consisting of the motion of the c.m. and 2. the part consisting of the motion of the particles in a frame of reference in which the c.m. is stationary. In this way, the kinetic energy of the earth is obtained by adding the contribution from its orbital motion to that from its rotational motion. The proof of this statement is very similar to the ones used earlier. It may be of help to some readers to repeat the arguments.

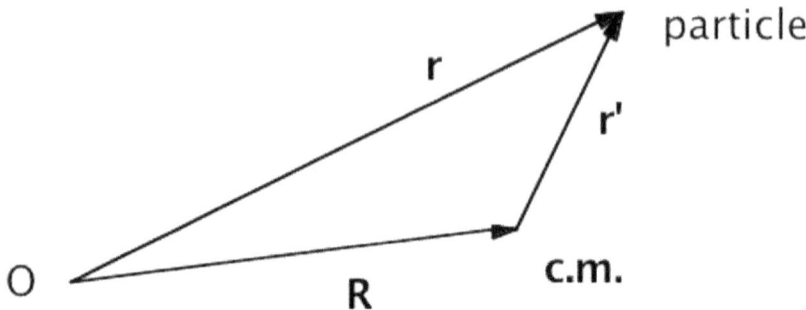

In the diagram, the point O corresponds to a stationary point in an inertial frame of reference. The position vector of some arbitrary particle can be regarded as the vector sum of the position vector of the c.m. and the position of the particle as measured from the c.m. All these vectors may be changing with time. The velocity vector of the particle is the rate of change of the position vector. It is obtained by differentiating $\mathbf{r}$ with respect to time, t. In this way, it can be expressed as the sum of the velocity of the c.m., $\mathbf{V}$, and the velocity of the particle with respect to the c.m., $\mathbf{v}'$. Note

$$\mathbf{r} = \mathbf{R} + \mathbf{r}',$$

and differentiate this equation with respect to time, leading to

$$\mathbf{v} = \mathbf{V} + \mathbf{v}'.$$

The kinetic energy of the particle involves the square of the speed. This is obtained by taking the scalar product of $\mathbf{v}$ with itself. The result is

$$v^2 = (\mathbf{V} + \mathbf{v}') \cdot (\mathbf{V} + \mathbf{v}')$$
$$= V^2 + v'^2 + 2\mathbf{V} \cdot \mathbf{v}'$$

When the kinetic energy of the system is calculated, the contributions from each particle are summed. The contribution from the first term in the equation is just $MV^2/2$, where M is the total mass. It treats the system as a particle moving with speed V. The contribution of the second term is the kinetic energy of the system as measured in a frame moving with the center of mass. The last term contributes nothing because the sum involves the scalar product of V with the total momentum, as measured with respect to the c.m. Since the c.m. is stationary in this frame, the result is 0.

Rotation about a Fixed Axis

Consider a flat disk, such as a CD, that is spinning about its symmetry axis with angular velocity ω. Its c.m. is on the axis and is therefore stationary. As shown earlier, the angular momentum is the same for every choice of origin. The calculation is simplest, however, if the origin is chosen at the c.m. The angular momentum of every particle is parallel to the angular velocity vector, since **r** and **v** are both perpendicular to this vector. (The disk is being treated as a two-dimensional object). The magnitude of the angular momentum of a particle in the disk is mrv or $mr^2\omega$. In order to obtain the total angular momentum, it is only necessary to sum the contributions from each particle. Since all the contributions are in the same direction, vector addition isn't necessary. It should be noticed that r is not only the distance to the c.m.—it is also the distance to the axis of rotation.

The summation procedure involving a continuous distribution of particles involves calculus. It is not important to be concerned with that now. You should merely note that the angular momentum is proportional to ω. The coefficient of proportionality is called the moment of inertia of the disk about its symmetry axis. It is symbolized by the letter I so that

$$L = I\omega.$$

The moment of inertia of a solid disk of mass M and radius R, such as a CD, about its symmetry axis is given by the formula

$$I = MR^2 / 2.$$

It turns out that the assumption that the disk is thin is unnecessary. That assumption was made so that it was easy to see that all the contributions were in the same direction. It will now be shown that the same equation holds for a solid circular cylinder about its symmetry axis. In order to see this, treat the cylinder as though it were made up of a huge stack of very thin disks, each of radius R. Its mass is just the sum of the masses of each disk. If you object to this conclusion on the grounds that this procedure involves the choice of different origins for different disks, you are overlooking something. The choice of origin was one of convenience. It has already been shown that this choice made no difference if the c.m. is stationary.

The examples used up to this point had rotational symmetry about the axis of rotation. That restriction isn't necessary. The arguments used so far would work equally well for many other shapes. For example, a similar analysis could be used for a rectangular parallelepiped. Such an object can be considered as being made up of playing cards that have been stacked together. As long as the axis of rotation is perpendicular to each card and passes through their centers, the same analysis works. In such situations, the angular momentum vector is in the same direction as the angular velocity vector.

In the most general situation the angular momentum vector need not be parallel to the angular velocity. An advanced treatment of this subject leads to an interesting conclusion that applies to all rigid bodies. It says that there always exist at least three perpendicular axes through the c.m. for which the angular momentum and the angular velocity are parallel. In the case of a cylinder, one of these axes has already been found. Any axis perpendicular to this also works. For a body with no symmetry, there are usually only three perpendicular axes that work. For a sphere, any axis works. This result is not surprising. However, it may be surprising that, as far as rotation is concerned, a cube has the same properties as a sphere.

The technique that was used to analyze the angular momentum of rotating cylinders can be used to calculate the kinetic energy of such bodies as well. In calculating the angular momentum the contribution from each particle is of the form $mr^2\omega$. The contribution to the kinetic energy is $mv^2/2$ or $mr^2\omega^2/2$. In these expressions, r is the particle's distance from the axis. When summed, this leads to the result

$$K = \frac{1}{2}I\omega^2.$$

For symmetrical rigid bodies rotating about a fixed axis, there is an analogy between rotation and the motion of particles in a straight line. The moment of inertia is analogous to the mass, the angular velocity is analogous to the velocity, and the torque is analogous to the force. In other words:

$$\tau = \frac{dL}{dt} = I\frac{d\omega}{dt} = I\alpha.$$

In this equation, α (the Greek letter *alpha*) is the angular acceleration. It is similar to F = ma, as applied to motion in one dimension.

Moments of Inertia

The moment of inertia of a rigid body, I, about some axis has been defined in terms of a sum. In some instances, it has to be determined experimentally. However, for rigid bodies of uniform density it is often possible to use integral calculus to calculate it. The moment of inertia associated with some common shapes is given below. It has been calculated for a symmetry axis through the c.m.

Thin-shelled hollow sphere of radius R: $I = \frac{2}{3}MR^2$

Solid sphere of radius R: $I = \frac{2}{5}MR^2$

Thin-shelled hollow cylinder of radius R: $I = MR^2$

Solid cylinder of radius R: $I = \frac{1}{2}MR^2$

Thin rod (baton) of length L: $I = \frac{1}{12}ML^2$

In the case of the thin rod, the axis of rotation is assumed to be perpendicular to the rod. There is a simple rule for calculating the moment of inertia of a body about an axis that doesn't pass through the c.m. It is known as the parallel axis theorem. If the moment of inertia about an axis through the c.m. is I_0, the moment of inertia about a parallel axis a distance d away is given by the rule

$$I = I_0 + md^2.$$

The symmetry of a sphere can be exploited to calculate the first moment of inertia on the list. Assume Cartesian coordinates are used with the origin at the center of the sphere. All the particles on the sphere are at the same distance, R, from the origin. Thus, if the Cartesian coordinates of some particle are (x, y, z), they satisfy the equation

$$x^2 + y^2 + z^2 = R^2.$$

Imagine a subscript i had been used to label each particle. If you multiply this equation by m_i, the mass of that particle, and sum, you find

The sum over i of $m_i x_i^2 + m_i y_i^2 + m_i z_i^2 = MR^2$,

where M is the total mass of the sphere. Because of the symmetry of a sphere, each term on the LHS contributes the same amount, namely one-third of MR^2. Now imagine the sphere to be rotating about the z-axis. The moment of inertia about this axis involves a similar sum, except the z-coordinate is absent. The moment of inertia is then two-thirds of MR^2, as shown above.

It is necessary to use calculus to calculate most moments of inertia. A short section devoted to calculus in which some useful integrals and derivatives are calculated is given at the end of this chapter. Those readers who wish to accept these results on faith are welcome to do so.

Motion without Constraints

This section deals with a sophisticated subject. It is designed to give the reader insights into the complex behavior of certain objects. Proofs will not be given. Consider the case of a football in flight. Neglecting air resistance, it is subject to no torques about its c.m. so that the angular momentum, **L**, about its c.m. is constant. If it is rotating about its symmetry axis, the angular momentum and the angular velocity are parallel to one another. In this case, the symmetry axis remains fixed, and there is no wobble in the ball's flight. Now consider the case where the spin axis makes an angle with the symmetry axis. Since the angular velocity, ω, behaves like a vector, it can be resolved into components. The component parallel to the axis leads to a component of **L** in the same direction. The component of ω perpendicular to this axis leads to a component of **L** in the perpendicular direction. Because the football is elongated along its symmetry axis, the moment of inertia in the perpendicular direction comes out to be larger than that along its symmetry axis. Even if the two components of angular velocity had been equal, the two components of angular momentum would be different. The perpendicular component would be larger. It follows that the angular momentum vector makes a larger angle with the axis of symmetry than the angular velocity vector does. Since the angular momentum is constant, both the symmetry axis and the angular velocity vector keep changing in direction.

The actual motion of a wobbling football is fairly complicated. However, the motion can be visualized by imagining two cones, one of which is rolling on the other. See the diagram.

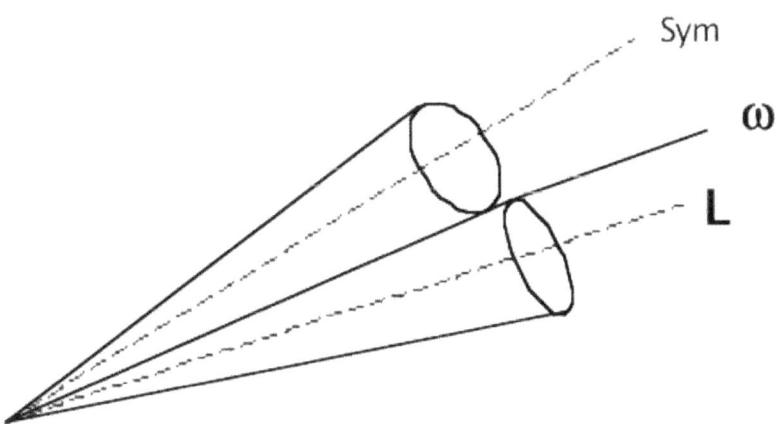

Imagine that the upper cone is fixed (embedded) within the football such that the lower left point is at the center of mass. The dotted line labeled "Sym" corresponds to the symmetry axis of the football. The angle of the cone is the angle made by the angular velocity vector with the symmetry axis. The orientation of the lower cone is fixed in space. Its angle is determined by the angle between **L** and ω. It really depends on the ratio of the two different moments of inertia involved. To visualize the wobble, imagine that the upper cone rolls over the lower cone, without slippage. What a spectator notices is the symmetry axis describing a cone in space. The angular velocity vector is also changing with time.

If the above ideas are applied to the earth, it is possible for ω to be changing even in the absence of torques. This could occur if the angular velocity vector is at some angle with the earth's axis of symmetry. The earth is not a sphere. It bulges in the equatorial region. Thus, the moment of inertia is larger in the symmetry direction than in the perpendicular direction. It follows that the angular momentum makes a smaller angle with the symmetry direction than the angular velocity does. In terms of the figure above, it is necessary to imagine the body-fixed cone as being inside the one that is fixed in space.

The most puzzling behavior of most spinning systems is how they behave when subject to external torques. When you first encountered a spinning top as a child, you probably wondered why it didn't fall over. Why does it seem to defy gravity? A similar situation is encountered when you ride a bike. At very slow speeds, the bike feels unstable. You have to be careful to keep your balance at these speeds. Once you've overcome your fear of speed, there is no problem staying vertical on a speeding bike. If you deliberately lean to one side, the bike might tend to turn, but it probably won't fall over. Examine the situation from the point of view of what happens to the angular momentum vector. The diagram below illustrates the ideas for the case of a spinning top:

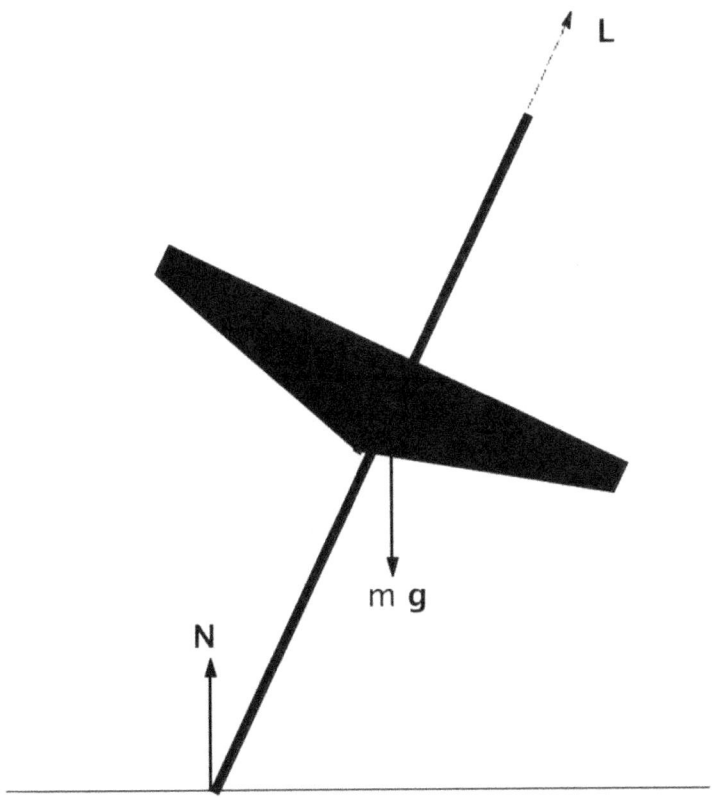

The external forces on the top are the gravitational pull of the earth and the contact force from below. If the table is smooth, this force is perpendicular to the surface. The vector N designates this force, while the vector **L** designates the angular momentum of the spinning top about the c.m. The reason for the letter N is because the word *normal* is a synonym for *perpendicular* in mathematics. If L is large enough, it is possible that N = mg so that the center of mass does not accelerate. This comes as a surprise to most people.

Consider the case for which there is a large spin, so that **L** is pointing along the axis of rotation, as shown in the diagram. The assumption that N = mg doesn't necessarily lead to a contradiction. There is a torque about the c.m. directed away from the reader (into the page). What does the angular momentum look like a moment later? It is necessary to add a small change to the vector that is already there. This can be accommodated without the length of the angular momentum vector changing. It merely changes its direction so that it is no longer in the plane of the page. The bigger L is the smaller the angle of turn. Thus, you can expect to see the axis of rotation gradually reorient itself without requiring any shift in the position of the c.m. This is a slightly oversimplified picture, as there is the possibility of a wobble.

Consider the opposite situation. In other words, imagine that the top were to fall over, even though it had an extremely large angular momentum. If the top were to end up horizontally, without changing its spin, the change in **L** would be huge. In order for the top to fall it would be necessary for N to be smaller than mg. There would be no force available to provide the required torque. Rapidly spinning tops can't just fall over.

An interesting lecture demonstration, given frequently in general physics courses, is to have a student stand on a turntable that is free to turn about a vertical axis with very little friction. The system is thus incapable of providing a torque with a component in the vertical direction. The student is handed a rotating bicycle wheel that has a convenient handle. The student holds it over her head like an umbrella. If the student tries to orient the axis horizontally the turntable starts spinning, so as to conserve the vertical component of angular momentum. The spectacle of a spinning student makes the audience roar.

There is the story of a clown who had mounted a large spinning wheel in a suitcase, driven by a quiet motor. He brought it with him when he registered at a ritzy hotel. This was in an era when hotels had bellhops who carried the luggage. Everything went well until the bellhop tried to round a corner. Try to imagine the rest.

It's Done with Mirrors

There seems to be a difference between the vectors that were introduced in this chapter and the ones introduced earlier. There is something artificial about the definition of a cross product, especially when a right-hand rule is used. Imagine that you are trying to communicate what *right-handed* means to beings on a distant planet. If their television sets scanned from right to left instead of left to right, pictures of our right hands would look like left hands. There doesn't seem to be any phenomenon, at the macroscopic level at least, that they could use to determine left from right. What does clockwise mean to someone who has never seen a clock? Try telling someone who had never seen a clock or a screw what a right-handed thread looks like.

There is a difference between vectors like τ, $\mathbf{L}$, ω and those like $\mathbf{F}$, $\mathbf{v}$, and $\mathbf{a}$. To distinguish between these types, the first ones are called *axial vectors*, or *pseudovectors*, while the others are called *polar vectors*.

To illustrate the difference, imagine that you are standing in front of a vertical mirror and are moving your hand to the right. You would say that the velocity vector of the hand as it moved pointed to the right. You would also say that the image in the mirror moved to the right. (You are not being asked what your mirror image would say. You should be focusing your attention on the image of the hand, not the image of the observer.) As long as the hand moves parallel to the plane of the mirror, the velocity vector is the same in the mirror as it is in the real world. However, when you move your hand away from you, and toward the mirror, the mirror image moves in the opposite direction. All polar vectors behave in this way. The component perpendicular to the plane of the mirror is reversed in direction. The other components remain unchanged.

Now consider the angular velocity vector. Standing in front of the mirror, turn your hand clockwise as though you were drawing a circle on the mirror itself. The mirror image is turning clockwise also. Thus, the angular velocity vector of the hand, and of its mirror image, is in the same direction, perpendicular to the mirror. Now imagine a situation in which the angular velocity vector is parallel to the mirror. It is not hard to see that polar vectors behave oppositely to those of axial vectors insofar as their mirror behavior is concerned.

You may wonder why anyone would be concerned about what happens in the mirror world. The answer to this question is not simple. It is related to the symmetry of physical laws. If you were to film almost anything that happens in nature through a mirror, it is unlikely that a viewer would realize that a mirror had been used. Newspapers and billboards might be unreadable, but these are human creations. It is only an accident of history that the English language is printed from left to right. Hebrew, after all, is printed from right to left. Certain mollusk shells are screw shaped with a right-handed thread. If these were filmed through a mirror, a biologist might recognize that a mirror was used, through the fact that the shells now appear left-handed. However, this doesn't mean that the left-handed thread violates some law of nature. It just means that somewhere in the evolutionary process something came along that favored the production of right-handed threads for this species.

Suppose you imagined a law of physics in which you could equate a vector to a pseudo-vector. In that case, if a phenomenon that obeyed this law were filmed through a mirror, the law would not be obeyed in the mirror world. It would be possible to recognize the fact that the filming took place through a mirror. The reason is simple. One of these quantities would be reversed in the mirror, while the other one wouldn't. Thus, vectors should be equated to vectors, while pseudo-vectors should be equated to pseudo-vectors. Although the cross product of two polar vectors is an axial vector, the cross product of an axial vector with a polar vector is a polar vector. An example of this is given by the equation

$$\omega \times \mathbf{r} = \mathbf{v}.$$

It is worth mentioning in this connection that physicists received a shock in the twentieth century when it was discovered that not all physical laws possess mirror symmetry. There is a type of phenomenon in nuclear physics, associated with the weak interaction, in which the mirror world does differ from the actual world. There is a way, after all, to distinguish right-handedness from left-handedness.

Mathematical Formulae

Readers are advised that they are free to skip this section, as nothing else in this book will depend on it. It is included for those readers who want to see how such quantities as moments of inertia can be calculated. These calculations involve integrals. It has already been emphasized that integrals are, in some sense, opposites of derivatives. This point becomes clearer by considering a few examples. A simple example deals with the derivation of the formula for the area of a circle of radius R, namely,

$$A = \pi R^2.$$

By definition, pi is the ratio of the circumference of a circle to its diameter so that

$$C = 2\pi R.$$

An integral is a type of sum. In order to think of the area of a circle as such a sum, imagine that it is divided into concentric thin circular strips. Use of polar coordinates allows you to designate a typical strip as extending between r and r + dr. If this strip were thin enough, you could imagine it to be a thin rod that had been bent into the shape of a circle. Its area would be approximately 2πrdr. The approximation becomes better as dr gets smaller. The area of a circle of radius R is the sum of the areas of all these strips. This sum is denoted by an integral and written as

$$A = \int_0^R 2\pi r dr.$$

What this equation says is that the area of the circle can be obtained by summing the areas of the strips starting from a strip near the center, where r = 0, and continuing until you get to R.

Consider what is meant by the derivative of A with respect to r. If A denotes the area of a circle of radius r, this derivative denotes the change in the area of the circle as its radius increases by dr. The change corresponds to the area of a circular strip of thickness dr and length 2πr. Division by dr and then taking the limit as dr approaches 0 leads to the equation

119

$$\frac{dA}{dr} = 2\pi r.$$

It's a common practice to evaluate an integral by guessing at a reasonable answer and checking the resulting expression by differentiation. If the derivative is correct, it is possible that the guess may differ from the correct answer by a constant. This results from the fact that the derivative of a constant is 0. However, it is usually easy to determine the constant. In order to see that πr^2 is the correct answer for the area of a circle of radius r, it is only necessary to differentiate r^2 with respect to r. This is the same type of derivative as was encountered when differentiating t^2 with respect to t. The derivative of πr^2 with respect to r is easily shown to be $2\pi r$. This must be the correct answer because the area is 0 when r is 0—no constant need be considered.

What would it mean if the lower limit were not 0 in the integral, just considered? In that case, the equation would appear as

$$A = \int_B^R 2\pi r dr = 2\pi \int_B^R r dr ,$$

This equation implies that you are computing an area within a region between a circle of radius B and radius R. An area, πB^2, has been removed from the original circle. The result is

$$A = \pi R^2 - \pi B^2.$$

Sometimes integrals are displayed in which no limits are specified. These are called indefinite integrals. The only thing that is desired in such a case is some expression for which the derivative gives the correct formula.

Exponents

It is helpful to discuss the use of exponents (the superscripts on algebraic quantities) before going further. The exponent 2 has already been used to denote the square of a quantity. Thus, x^2 means the product of x with itself. If this quantity is multiplied by x once more, the result is x cubed. This would be written as x³. The meaning of exponents, such as 2, 3, 4, should be clear. They satisfy the rule

$$x^m x^n = x^{m+n}.$$

This formula can be used to extend the definition to exponents like 0 and 1. To see this, note that the product x^n and x^0 is x^n, and multiplication of x^n by x^1 is x^{n+1}. This is equivalent to the following:

$$x^1 = x, \qquad x^0 = 1$$

The rule can now be used to define negative integers as exponents. Consider the following:

$$x^{-n} x^n = x^0 = 1, \text{ so that } \quad x^{-n} = \frac{1}{x^n}.$$

Fractional exponents such as 1/2 can be defined by the rule

$$x^{1/2} x^{1/2} = x^1 = x, \quad \text{so that } x^{1/2} = \sqrt{x}$$

The definition can be extended even further but is not needed.

In order to derive the formulas for the moments of inertia given earlier in this chapter, it will be necessary to derive some formulas for differentiating algebraic expressions first. The following formulas have already been derived:

$$\frac{dc}{dx} = 0, \quad \frac{d(cx)}{dx} = c, \quad \frac{d(cx^2)}{dx} = 2cx,$$

where c is any constant. It turns out that this list can be extended to any integer power of x. It will be found that

$$\frac{dx^3}{dx} = 3x^2, \quad \frac{dx^4}{dx} = 4x^3, \quad \frac{dx^5}{dx} = 5x^4, \text{............} \frac{dx^n}{dx} = nx^{n-1}$$

The symbol n in this expression stands for any integer whatsoever. (It actually works for any number, but it will only be proven for integers.) The proof is one of induction. This means that it is assumed to be true for some integer n. That assumption will lead to the conclusion that it is also true for the next higher integer. The reasoning goes as follows: It is already known that it is true for n = 2. That assumption makes it true for n = 3. This in turn makes it true for n = 4. Continuing in this way leads to the conclusion that it is true for all integers.

To carry out this inductive proof, recall the formula

$$\frac{dfg}{dx} = f\frac{dg}{dx} + g\frac{df}{dx},$$

which holds for arbitrary functions, f and g. Now use

$$f = x^n, \quad g = x, \quad \text{so that } fg = x^{n+1}.$$

It is found that

$$\frac{dx^{n+1}}{dx} = x^n + xnx^{n-1} = x^n + nx^n = (n+1)x^n.$$

This completes the proof.

The reader is encouraged to prove that

$$\frac{d(1/x)}{dx} = \frac{dx^{-1}}{dx} = -1x^{-2} = -\frac{1}{x^2}$$

Hint: Let f = x, g = 1/x.

You might also try differentiating the square root of x by choosing this for both f and g. You should find that the formula for differentiating x^n that was given earlier works for all values of n, not just positive and negative integers.

Expressions for the surface area and volume of a sphere are needed in order to calculate the moment of inertia of a solid sphere. For this purpose, restrict your attention to the earth and assume it is a perfect sphere. It will be necessary to find a more convenient position coordinate than latitude. Imagine a straight line that runs along the axis of rotation from the South Pole to the North Pole. Imagine this line to be the x-axis of a Cartesian coordinate system, in which the origin is at the earth's center. Now focus your attention on a particular meridian. It forms a semicircular arc. The y-coordinate of a point on this meridian is the distance of this point from the axis of rotation. The position of any point on this meridian can be specified by means of polar coordinates also. The North Pole has an angular coordinate of 0, a point on the equator has an angular coordinate of $\pi/2$ radians, and the South Pole has an angular coordinate of π radians. All these points have the same radial coordinate, namely, R, the radius of the earth. The position of any point on the earth can be specified by means of these polar coordinates along with the longitude.

In order to calculate the surface area of the earth, it is useful to divide the surface into thin circular strips parallel to lines of latitude. This is similar to the procedure used earlier in calculating the moment of inertia of a thin disk. In the present case, the radius of each strip is given by the formula

$$\text{rad} = R\sin\theta,$$

so that

$$\text{length} = 2\pi R\sin\theta, \quad \text{width} = Rd\theta.$$

The total area of the earth is then given by an integral, namely,

$$A = 2\pi R^2 \int_0^\pi \sin\theta \, d\theta = -2\pi R^2[\cos\pi - \cos 0]$$

$$= -2\pi R^2(-1-1) = 4\pi R^2.$$

Use has been made of the fact that the derivative of the cosine is the sine, except for a minus sign. Thus, this integral results in the negative of the cosine.

The volume of a sphere is obtained by making use of the fact that the volume of a thin shell is the product of the area with the thickness. By dividing a sphere into a system of thin shells, you are led to the following integral:

$$A = 4\pi \int_0^R r^2 dr = 4\pi R^3 / 3.$$

The result of this calculation can be checked by differentiation.

In order to determine the moment of inertia of a solid sphere, it is useful to divide it up again into thin shells. The moment of inertia of a shell has been shown to be given by the product of the mass with two-thirds of the square of the radius. The mass is expressed in terms of its density, ρ, and volume. The contribution to I from such a shell is thus:

$$\frac{2}{3} 4\pi r^2 dr \rho r^2 = \frac{8\pi\rho}{3} r^4 dr.$$

The integral of r to the fourth power is r to the fifth power divided by five. In order to see this, you should check by differentiation. In order to express the final result in terms of the mass, it helps to recall that the mass is the product of the volume of the sphere with its density. The final result can be written as

$$I = \frac{2}{5} MR^2.$$

Chapter 7
Applications

The theory developed in the preceding chapters is useful in dealing with the motion of some simple systems. Examination of these systems leads to a deeper understanding of the theory. For a particular example, consider the periodic motion of a physical pendulum, one that is made out of a solid material instead of a string. Such a pendulum is shown in the figure below.

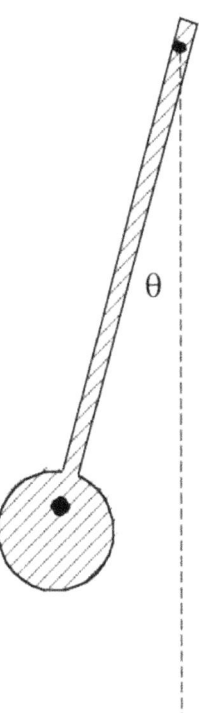

The small black circle indicates the axis about which the pendulum swings. The larger circle is the center of mass. Let ℓ be the distance between these two points. The torque about the axis is in a direction perpendicular to the diagram. You should be able to verify that its magnitude is $mg\,\ell \sin\theta$. If I is the moment of inertia of the pendulum about this axis, it follows that

$$mg\ell \sin\theta = -I\alpha.$$

The minus sign indicates that the angular acceleration is in the direction of decreasing θ. It should be noted that the gravitational force, mg, behaves as though all the mass were concentrated at the c.m. This statement will be proved later in this chapter.

It is difficult to analyze this problem if the swing of the pendulum is large. Note that $\ell \sin\theta$ is just the distance of the c.m. from the axis. On the other hand, if the angle is measured in radians, $\ell\theta$ is the size of the circular arc the CM has traveled past the bottom. For small angles, these two quantities are very nearly equal. In other words, the following approximation is valid for small oscillations:

$$\alpha = -\frac{mg\ell}{I}\theta.$$

You are encouraged to ask yourself what previous relationship this resembles. It is exactly analogous to the equation for SHM, in which angular quantities replace linear ones. In other words, θ replaces x, and the angular acceleration replaces the linear acceleration. Thus, for small oscillations, the frequency of the pendulum doesn't depend on amplitude. The period of the oscillation is given by the relation

$$T = 2\pi\sqrt{\frac{I}{mg\ell}}.$$

The numerator in this equation depends on ℓ. The moment of inertia about the axis is given by the parallel axis theorem, namely,

$$I = I_0 + m\ell^2.$$

Here I_0 denotes the moment of inertia about the c.m.. Although the formula for the period seems to depend on mass, this is not the case, since the moment of inertia is proportional to the mass.

It is useful to see whether the formula for the period fits the results already obtained for the simple pendulum. In that case, the parallel axis theorem yields $I=m\ell^2$. The period is the same as the one found earlier, namely,

$$T = 2\pi\sqrt{\ell/g}.$$

This formula for the period of a simple pendulum is probably well-known to every graduate student in the field of physics. It brings to mind a true story about a question asked of such a student on a test. It was part of the candidacy exam given to graduate students seeking a doctoral degree. In this exam a student stood in front of a blackboard while a committee of about five professors bombarded the candidate with questions. One of the professors asked the student whether he was familiar with the formula for the period of a simple pendulum. After being assured that he was, the professor then asked, "What's the period of a pendulum of infinite length? You can ignore the fact that the earth is rotating." The student had the good sense to realize that he was being set up. It had all the earmarks of a trick question. The formula suggested that the period would be infinite. The obvious thought was that the bob was moving along a straight line parallel to the surface of the earth. Why should it swing at all? Luckily the student had the correct thought. Aha! The earth is a sphere, and the bob is moving in a straight line tangent to the sphere. In deriving the equation for the period of a pendulum, almost everyone neglects the fact that the earth is round. The student could see that the flat motion with respect to the curved earth of the infinite pendulum was equivalent to the curved motion of a finite pendulum with respect to a flat earth. The length of this finite pendulum would equal the earth's radius. The student then realized that the resulting formula was the same as the one for the period of a satellite in low orbit around the earth. The answer he gave was "about an hour and a half." There were smiles all around. The student passed the exam.

Problem

Consider a pendulum made by passing a needle through a uniform disk at a point halfway out from the center. The needle serves as a horizontal axis about which the disc rotates. What is the period of such a pendulum? (You may use the formulas of the previous chapter to determine the moment of inertia about this axis.) Show that you get the same period if the needle is placed at the rim. Where can you stick the needle to get the shortest period? (Hint: The last question can be answered by finding the formula for the dependence of the period on the distance of the needle from the center. When the period is a minimum, the derivative of this expression with respect to the distance is 0. If the derivative is positive, for example, the period can be reduced by making the radius smaller.) The answer is discussed at the end of the chapter.

Rolling

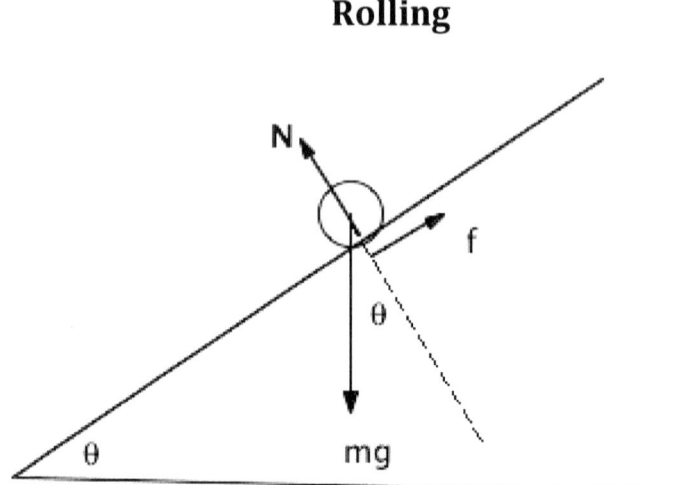

Consider a billiard ball with radius R that is rolling down an inclined plane as shown in the figure above. The contact force exerted by the plane is broken into two components, tangential and normal. The tangential component is f, the frictional force. It is shown below the inclined plane for clarity. It is actually along the plane. Assuming there is no slippage, the speed of the c.m. is given by ωR, as can be seen from the following argument. This is the speed that the point of contact is moving with respect to the center. Since there is no slippage, it is the speed of the center with respect to the plane.

Differentiating the expression $v = \omega R$ with respect to time, it is found that the acceleration of the c.m. is related to the angular acceleration. It is

$$a = \alpha R.$$

The problem is to find the magnitude of the acceleration. Some textbooks deal with this problem by taking torques about the point of contact of the sphere with the incline. It has the advantage of not involving the unknown frictional force. However, as the point of contact is not the c.m., nor is it fixed with respect to an inertial frame, this procedure is suspect. It is safer to take torques about an axis through the c.m., noting that f is the only force that exerts a torque about this axis. This leads to the equation

$$fR = I\alpha$$

or, equivalently,

$$fR^2 = Ia.$$

It has been shown that the moment of inertia of a homogeneous solid sphere is given by $I = \frac{2}{5}mR^2$ so that $f = \frac{2}{5}ma$. Application of Newton's second law leads to

$$mg\sin\theta - f = ma$$

$$mg\sin\theta = \frac{7}{5}ma$$

$$a = \frac{5}{7}g\sin\theta.$$

From this result, it is possible to determine all the forces on the sphere. The net or resultant force is down the plane and of magnitude $5mg\sin\theta/7$, and the frictional force is $2mg\sin\theta/7$.

Is energy lost because of rolling friction? In order to answer this question, imagine that the sphere rolls down a distance, D, starting from rest. The c.m. behaves like a particle. The work done on this particle by the constant resultant force is $5mgD\sin\theta/7$. The result of this calculation leads to the following equation:

$$\frac{1}{2}mv^2 = 5mgD\sin\theta/7.$$

It is necessary to add the kinetic energy due to rotation to find the total energy gained. It turns out to be two-fifths of the result just obtained, because

$$\frac{2}{5}(\frac{1}{2}mv^2) = \frac{1}{2}(\frac{2}{5}mR^2\omega^2) = \frac{1}{2}I\omega^2.$$

Addition of the two contributions to the kinetic energy leads to the result

$$K = mgD\sin\theta.$$

This equals the loss in gravitational potential energy. It seems that energy is conserved when an object rolls without slipping.

Do you wonder how energy can be conserved in the presence of friction? The situation must have been idealized somehow. Yes. The idealization comes from the fact that it is assumed that the plane and the sphere are so hard that they only make contact at a point. In any real situation, contact is made over a finite area. It is easily noticed that balls don't roll very far on a soft material, such as a carpet.

An Interesting Problem

Suppose a bowling ball is thrown in such a way that it is not spinning at all. For a moment, it just slides along the lane at a speed, v. Treat the ball as a uniform solid sphere of known mass, m, and radius, R. Because of friction, there will be a net force on the ball and also a torque about its c.m. Eventually the slippage will cease, and the ball will roll as in the preceding example. It turns out that the speed and the angular velocity can be determined when this happens. It is not necessary to know how much sliding friction there is. If you'd like to try finding these quantities, consider the following suggestions. Draw a diagram showing all the forces acting on the bowling ball. Note that there is at least one choice of origin about which the total torque is 0. Show that the angular momentum about this origin is conserved and use that information to solve the problem. You might want to review the section involving the center of mass.

A Puzzling Situation

A man goes to a carnival and encounters a barker who is offering a challenge involving a new piece of apparatus. The apparatus is similar to a merry-go-round. It contains a track along the perimeter on which a person can walk. (See the diagram below.) There are no motors driving the apparatus. It is extremely light. Nevertheless, it is strong and mounted on excellent bearings. The barker invites the man to enter through a gate in front and walk halfway around

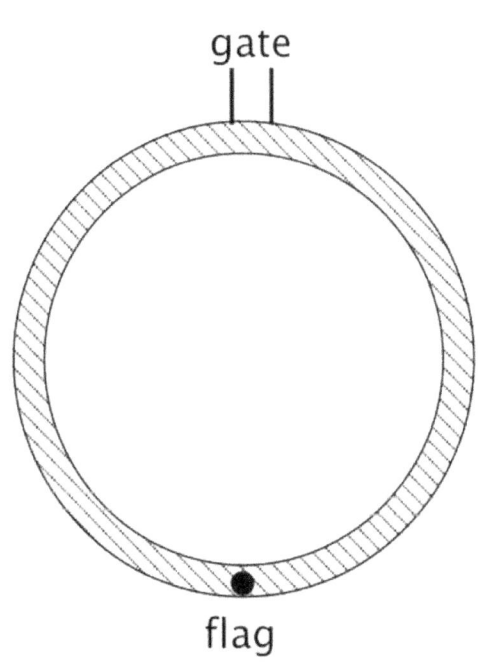

gate

flag

on the track to a flag mounted at the back end of the merry-go-round. The man is told that the apparatus will be locked in position while he walks to the back and that it will be free to turn after that. The admission fee to the apparatus is ten dollars, but there is a fifty-dollar prize if the man can find his way to the front gate (while staying on the track) within five minutes, with the following catch: The merry-go-round is free to turn once he starts walking, and he must get to the front gate before the flag gets there.

Being tempted by the fifty-dollar prize, the man accepts the challenge. He pays the ten dollars and walks to the flag. As promised, the apparatus is locked. At this point the man decides to walk clockwise along the path. He walks about fifteen feet along the path but notices that he only travels about five feet with respect to the ground. The flag has traveled about ten feet from its original position, in the counterclockwise sense. The man realizes that the apparatus is much lighter than he anticipated. Its moment of inertia about its axis of rotation is only about half of his own. He remembers that angular momentum is conserved in the absence of torque. Whatever angular speed he attains, it will only be half that of the apparatus. He thinks a while longer. Perhaps if he takes off a shoe and throws it backward, the force that the shoe exerts on him during the throw will be enough to get the whole apparatus to turn clockwise. He rejects the thought. His shoes are worth more than fifty dollars. Perhaps he can suck in air while facing forward and blow it out in the backward direction. He tries it and quickly realizes that even if the bearings were perfect, he couldn't succeed in the time allowed.

Question: Is the situation hopeless? What might you try if you were in this situation? See the discussion at the end of the chapter.

Collisions

When a baseball and bat collide, an interesting sequence of events take place. A bat is not simply a rigid body in which all particles keep a fixed distance from one another. Rather, a sudden deformation takes place at the point(s) of contact, and vibrations travel down the length of the bat at the velocity of sound in wood. Vibrations at the handle are not felt instantaneously since it takes some time for these waves to propagate down to it. When a batter strikes at the ball, perhaps hoping for a home run, he would like the part of the bat that meets the ball to maximize the transfer of momentum, but not at the expense of feeling too much pain in his hands. This is not a trivial problem, and research continues on it into the twenty-first century. However, you can consider a similar problem and see where an elementary analysis takes you.

In the figure below, the vertical line specifies a narrow bar of mass, M, and length, L, that is about to be struck by a ball of mass, m. The ball hits it at a distance, h, above its c.m.

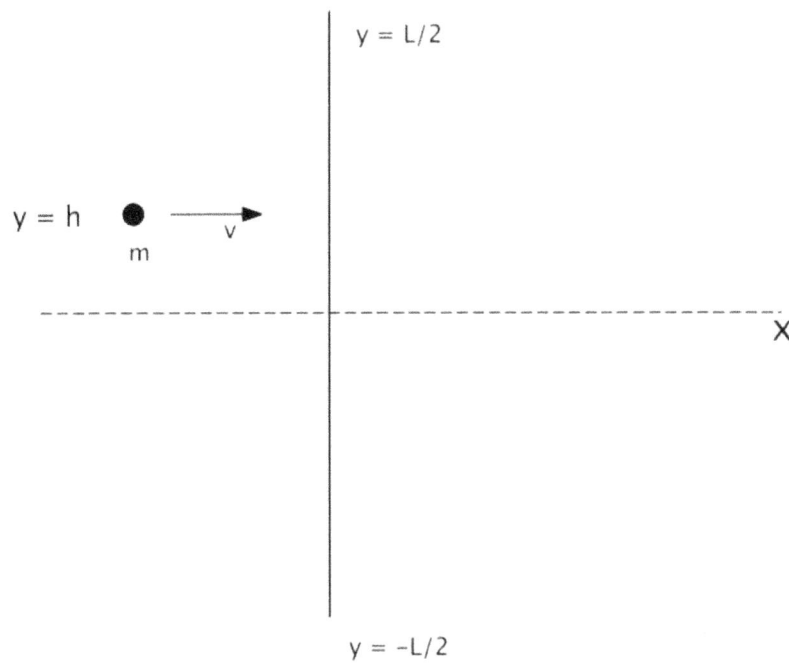

Suppose that the magnitude of the momentum transferred in the collision is given by the symbol P. You should be able to deduce that the angular momentum transferred (about the c.m.) is Ph. From this information, it can be concluded that the velocity, V, of the c.m. after impact is given by

$$V = \frac{P}{M},$$

and the angular velocity is given by

$$\frac{Ph}{I} = \frac{12Ph}{ML^2}.$$

(See the previous chapter for the moment of inertia of a rod.) Try to answer the following question: Is there any value of h for which the velocity of the bottom point is 0 just after impact? In such a case, you could be holding the bar at the bottom and not be hurt by the impact.

From knowledge of the angular velocity, you can determine the velocity of the bottom with respect to the c.m. Once that is known, it is a simple matter to determine its velocity with respect to the ground. This velocity is found to be

$$v' = \frac{P}{M} - \frac{12Ph}{ML^2}(L/2) = \frac{P}{M}(1 - \frac{6h}{L}).$$

This can be made to be 0 if the ball strikes when h =L/6. The central idea is that there is some point of impact that results in a minimum amount of discomfort. Tennis players become aware of a "sweet spot" in their rackets.

Center of Gravity

It has been assumed that the gravitational force on a system of particles acts as though it passes through its center of mass. This implies that the torque produced by using this assumption is the sum of the contributions coming from each particle. In fact, the center of mass of such a system is often called the center of gravity. Show that this is exactly valid in the case of a flat earth, which means that the vector **g** is constant. Hint: Choose the direction of the gravitational force to be along the y-axis of a Cartesian coordinate system.

Answers

Pendulum in the Form of a Disk

The formula that was derived for the period of a pendulum is

$$T = 2\pi \sqrt{\frac{I}{mg\ell}} \; .$$

Use of the parallel axis theorem for I yields

$$T = 2\pi \sqrt{\frac{mR^2/2 + m\ell^2}{mg\ell}} = 2\pi \sqrt{\frac{R^2/2\ell + \ell}{g}} \; .$$

Consider the following two cases:

1. $\ell = \dfrac{R}{2},$ 2. $\ell = R$

$$T = 2\pi \sqrt{\frac{R^2/R + R/2}{g}} = 2\pi \sqrt{\frac{3R}{2g}} \; .$$

2.

$$T = 2\pi \sqrt{\frac{R/2 + R}{g}} = 2\pi \sqrt{\frac{3R}{2g}} \; .$$

The two situations lead to the same period.

The expression for the period depends on only one variable. As explained in the hint, the derivative of the expression with respect to that variable should be 0 at a maximum or a minimum. It is not necessary to deal with the entire expression. The period will be a minimum when the expression in the numerator within the square root is a minimum. A more familiar symbol will be used for the variable. Thus, the expression to be minimized is of the form

$$\frac{R^2}{2}x^{-1}+x.$$

It has already been shown that

$$\frac{dx^n}{dx}=nx^{n-1}.$$

The minimization process leads to

$$-\frac{R^2}{2}x^{-2}+1=-\frac{R^2}{2x^2}+1=0.$$

Returning to the original notation, the result is

$$\ell=\frac{R}{\sqrt{2}}=.707\,R.$$

The minimum period is

$$T_{min}=\frac{1}{2\pi}\sqrt{\frac{\sqrt{2}R}{g}}\cong\frac{1}{2\pi}\sqrt{\frac{1.414R}{g}}.$$

This period is only slightly shorter than the period when the disk is suspended at its rim.

Solution to an Interesting Problem

The angular momentum of the bowling ball in this problem is the vector sum of two contributions: 1. the angular momentum due to rotation and 2. the angular momentum associated with the motion of the c.m. Suppose you choose an origin on the alley directly in front of the ball. Initially the magnitude of its angular momentum about this point is mvR. (If you have trouble with this, review the discussion of cross products. Only the vertical component of the position vector is involved.) If you were to draw a force diagram, it would be obvious that the normal force on the ball is balanced by the weight force (since there is no vertical component of acceleration). The only other force is that of sliding friction. This force contributes nothing to the torque about a point on the surface of the alley dead ahead of the ball. Thus, the angular momentum about this point is conserved. When the ball starts rolling, its angular velocity is given by V/R, where V is its new velocity. The new expression for the angular momentum is

$$mVR + I\omega = mVR + \frac{2}{5}mR^2\frac{V}{R} = \frac{7}{5}mVR.$$

Equating this to the original angular momentum yields

$$V = \frac{5v}{7}.$$

The interested reader should be able to show how much energy has been lost. The answer is that it loses two-sevenths of its initial kinetic energy. This problem is interesting in that nothing seems to depend on the magnitude of the frictional force. Of course, the time and place where rolling starts depends on this force.

Solution to a Puzzling Situation

The man should recall that if he obtains some angular momentum about a vertical axis, then the rest of the apparatus will gain an equal and opposite angular momentum. It is only necessary to spin while standing still with respect to the apparatus. It would be useful to spread out his arms and possibly his legs so as to increase his moment of inertia. If he spins counterclockwise in place the whole apparatus will turn clockwise. After the apparatus has turned through a little more than 60^0 he can then walk the rest of the way. He will be able to do this because the flag is over 240^0 from the gate, while he has less than 120^0 to go. The reader may wonder where the torque on the apparatus is coming from. The answer is that when the man starts to spin, he must get a torque from the apparatus. The reactions to the forces that the apparatus exerts on his feet while he starts to turn will help resolve these questions.

Center of Gravity Solution

The diagram below shows a system of five particles. The gravitational force on particle #1 is shown. The gravitational forces on all the other particles are given by similar expressions, and they are pointing in the same direction. It is also assumed that the acceleration due to gravity, g, is constant. The masses need not lie in the plane of the diagram. The third Cartesian coordinate, z, is perpendicular to the page. Recall that the z component of the cross product $\mathbf{A} \times \mathbf{B}$ is given by $A_x B_y - A_y B_x$. For the time being just focus on this component of the torque. For particle #1, it is given by

$$\tau_{1z} = -x_1 m_1 g = -g m_1 x_1.$$

The total torque is then given by the equation

$$\tau = -g(m_1 x_1 + m_2 x_2 + m_3 x_3 + \cdots) = -gMX,$$

where M is the total mass, and X is the x coordinate of the c.m. This implies that the resultant of the gravitational forces, namely, Mg, behaves as though it passes through the x coordinate of the c.m. A similar analysis can be given for the z-coordinate.

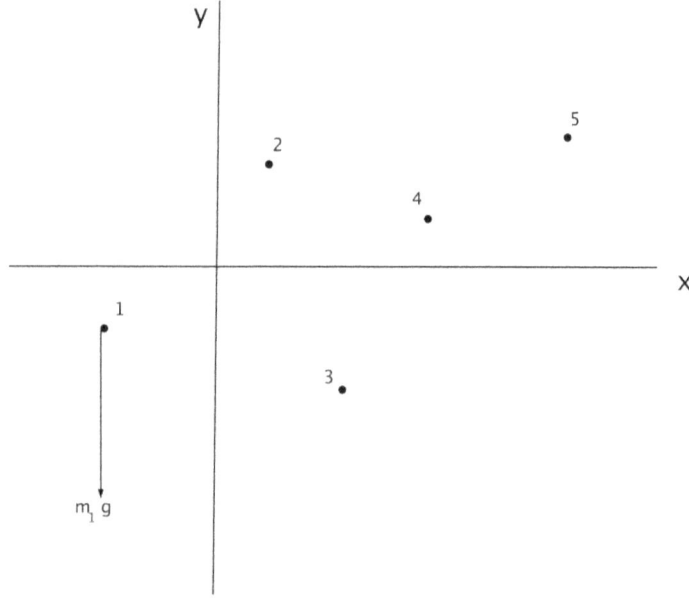

Chapter 8
Waves on Strings

There are many types of waves. Sound waves, light waves, water waves are familiar examples. The simplest wave to imagine is that which takes place on a thin string under tension. This is the type of motion that takes place on a string of a musical instrument, such as a violin or a guitar. The problem with the analysis of strings on musical instruments is that they have ends. It is easier to deal with strings that have no boundaries and put in the boundary conditions later. For this purpose, imagine a horizontal infinite string that runs along the x-axis of a Cartesian coordinate system. Consider an infinitesimally small section of this string that extends from x to x + dx. Newton's second law will be applied to this infinitesimal section. The only important forces that need be considered are from the adjacent section of the string. Gravity is unimportant and will be neglected.

The following assumptions are made: 1. The string undergoes only small vibrations so that it is nearly horizontal everywhere. 2. The motion of the particles is vertical along the y-axis. One conclusion that follows from this set of assumptions is that the horizontal components of the force at each end cancel. There is no acceleration in the x direction. This implies that the tension in the string doesn't depend on position. In other words, the magnitude of the force that one section of string exerts on its neighboring section is the same throughout the string. Don't worry about the fact that the string is not exactly horizontal as it vibrates. This contributes negligible error to the assumption that the x-component of force is T, for small vibrations.

You may ask how there could be any vertical force if the tension doesn't vary with position. The answer has to do with the fact that the string doesn't remain exactly horizontal. The y-component of the force at some particular instant comes from the fact that the string makes an angle with the horizontal. This angle depends on the derivative of y with respect to x. It is necessary to make a change of notation in specifying such a derivative. In the previous examples in which derivatives were used, the quantity being differentiated depended on only one variable. In the present case, the variable y depends on two variables, namely x and t. It is necessary to use a notation such that you know a variable is being held constant while the other variable is changing. The following notation is used to show that the slope of the string is being considered at one instant of time:

$$slope = \frac{\partial y}{\partial x}.$$

In a similar fashion, the velocity of a particle in the y direction at a given value of x would be written as

$$v = \frac{\partial y}{\partial t}.$$

These are called partial derivatives. The Greek letter *delta* signifies that all other variables are being held constant, except for the one in the denominator.

In order to obtain the acceleration of the particle at a given position, it is necessary to take another time derivative. In order to save writing, an abbreviation has been adopted, as follows:

$$a_y = \frac{\partial}{\partial t}(\frac{\partial y}{\partial t}) = \frac{\partial^2 y}{\partial t^2}.$$

You should not let the notation be a source of confusion. Nothing is being squared. This equation may be read as "The acceleration is the second derivative of y with respect to t." If another time derivative is needed the number 2 will be replaced by the number 3. Third derivatives do not occur frequently in physics.

Consider, once more, the net force on the section of string between x and x + dx. The right end of the string is being pulled in the y direction by the adjacent portion of the string. The rigorous expression for this component of force is

$$F_y = T \sin\theta,$$

where theta is the angle the string makes with the horizontal at x + dx.

Since only very small angles are involved, it is possible to replace the sine by the tangent. If this is not clear, try to visualize a right triangle with one small angle and compare the sine of this angle with the tangent. Since the tangent is equivalent to the derivative of y with respect to x, this specifies the force as

$$F_y = T \frac{\partial y}{\partial x}.$$

This force is in the positive direction when the derivative is evaluated at x + dx. It is in the negative direction at the other end of this infinitesimal region. The net force in the y direction is the difference of these two terms. It can be considered as being due to the product of the rate of change of F_y with dx. The net force in the y direction is thus

$$T \frac{\partial^2 y}{\partial x^2} dx.$$

We are in need of one more thing, the mass of this section of string. The mass depends only on the length of this section, namely, dx, and the mass per unit length, which is usually denoted by the Greek letter mu. Newton's second law leads to

$$T \frac{\partial^2 y}{\partial x^2} dx = \mu dx \frac{\partial^2 y}{\partial t^2}.$$

As long as the string is nearly horizontal, the approximations that are made become accurate as dx approaches zero. In any case, dx factors out of the equation. So, for small oscillations, the resulting equation is

$$\frac{\partial^2 y}{\partial x^2} - \frac{1}{c^2} \frac{\partial^2 y}{\partial t^2} = 0, \quad \text{where } c = \sqrt{T/\mu}.$$

The reason for introducing the constant c is because this constant turns out to be the speed of wave propagation on the string. This will be proven shortly. This is a famous equation, known as the wave equation. Its solutions have interesting properties.

Many new ideas have been introduced in deriving this equation. It takes time to absorb them all. The first term in the equation is the rate at which the slope changes as you move along the string at any instant. It is 0 when the section of string is straight. It is thus a measure of the curvature of the string. Except for a constant, the second term is the transverse acceleration of a point on the string. There is no acceleration for a particle on a straight section of the string.

There are an infinite number of solutions to the wave equation. It will be seen that the most general solution for the infinite string is a superposition of two waves traveling in opposite directions at the speed c. The idea of superposition is clarified in the next section.

The Principle of Superposition

The following statements are to be proved: 1. If a solution has been found in the form $y = f(x, t)$, it can be scaled up or down (that is, multiplied by a constant), and it will still be a solution. In other words, like a harmonic oscillator, the amplitude of the vibration is arbitrary. It is necessary to put in a warning. The oscillations must be small enough so that the string is almost always nearly parallel to the x axis. 2. If there exists a second solution to the equation, $y = g(x, t)$, then these solutions can be added so that $y = f(x, t) + g(x, t)$, yielding still another solution. It is necessary to make the same warning as before.

The two statements constitute the principle of superposition. In order to prove this principle, you should recall that the derivative of a sum is the sum of the derivatives. You should think of the process of differentiating a function as an operation whereby something is done to a function of x and t in order to get a new function. Another operation that you may consider is the process of squaring. You should note that the square of a sum is not necessarily the sum of the squares. This is illustrated by the following:

$$(f+g)^2 = f^2 + g^2 + 2fg.$$

Such an operation is said to be nonlinear. A linear operator has a special property. If f and g are functions of x and t, and A and B are constants, an operator, LIN, is linear if you can conclude that

$$LIN(Af + Bg) = ALINf + BLINg.$$

The wave equation is equivalent to the statement that the result of applying a linear operator to some particular function is zero. If a second function can be found that also yields 0, it should be clear that the superposition of these functions also yields 0.

One reason the idea of superposition is useful is that it tells us that a string can vibrate in SHM (simple harmonic motion) with several different frequencies at the same time. If not, a violin would sound like a tuning fork. The superposition principle is a much more powerful idea than that, however. It applies to light waves as well as to waves on strings. Try to imagine a world in which the principle is not valid. Suppose, for example, you are looking at a scene in which you see flowers and a lake and a mountain. The fact that you can distinguish the flowers, the mountain, and the lake is because the human eye and the human brain can decompose the superposition of signals into their individual components. If the light waves coming from the flowers were to combine with the waves from the lake in some more complex way than a simple superposition, it would result in a jumble that the brain couldn't disentangle. The same is true of sound waves. The fact that you can distinguish the different instruments while a symphony is being played is a consequence of linearity. A medium may behave in a linear fashion only for waves of sufficiently small amplitude. A shock wave, like that of an explosion, is an example of nonlinearity, for sound waves. Various waves may contribute to the phenomenon, but the result is usually one loud bang. A world without linearity would be completely chaotic.

Traveling Waves

In order to understand the mathematical description of a wave on a string, it helps to deal with a bit of mathematics first. Imagine that y is some continuous and smooth function of only the position variable, so that y = f(x). This implies that if you choose some numerical value of x, there is a precise formula for obtaining the value of y at that value of x. Suppose, for example, that feet has been chosen as the unit for x, and millimeters for the unit of y, such that f(2) = 7. This implies that at a point 2 feet to the right of some chosen origin, the string is 7 millimeters above its equilibrium value. The question is now put to you: "What does f(x - 6) look like?" The answer is: "It would look exactly like the previous shape, except that it has been shifted 6 feet in the positive x direction." In order to illustrate this, it will be shown that at a point 6 feet to the right of our previous point, the displacement is 7mm from its equilibrium point. Thus, we look at the new function at x = 8. It is found that f(x - 6) = f(8 - 6) = f(2). This is 7 mm, according to the formula. It has the same value that the point at x = 2 had, according to the original equation. It is easily seen that f(x -A), where A is some numerical constant, is the original function (or shape) except that it has been shifted A feet in the direction of increasing x.

It will now be shown that solutions to the wave equation exist in which some arbitrary shape (except that it has to be small, continuous, and smooth) travels down the string with constant speed in either direction. The speed is determined only by the properties of the string, namely, the mass per unit length and the tension. Translated into an equation, this statement says

$$y = f(x - ct)$$

is an acceptable solution to the wave equation. In order to understand this, simply note that at any particular instant of time, the solution has the form f(x - A). The only difference with the previous solution is that A is changing at the rate c. The constant c has the units of a velocity in feet per second when x is measured in feet and t is measured in seconds. If the propagation is in the direction of decreasing x, c is negative.

In order to show that the assumed solution works, it is useful to define a new variable by the following equation:

$$x' = x - ct.$$

The assumed solution can be written as f(x'). It is a function of a single variable, although x' is a function of two variables. Recall that derivatives behave like fractions. In this way it can be seen that

$$\frac{\partial f}{\partial x} = \frac{df}{dx'}\frac{\partial x'}{\partial x} = \frac{df}{dx'},$$

$$\frac{\partial f}{\partial t} = \frac{df}{dx'}\frac{\partial x'}{\partial t} = -c\frac{df}{dx'}.$$

Note that partial derivatives are only needed when some quantity is being held constant during the differentiation process. It should be clear that if another derivative is taken in order to get second derivatives, another factor of -c is introduced into the time derivative. It is found that

$$\frac{\partial^2 f}{\partial t^2} = c^2 \frac{\partial^2 f}{\partial x^2}.$$

This is the wave equation, in a slightly different form from the one shown earlier. This proves that the assumed solution of a traveling wave works.

Stringed Instruments

No treatment of waves on strings would be complete unless it included the vibrations of stringed instruments. It is thus necessary to include the boundary conditions of strings with fixed ends. That means that the value of y must be zero at each end. Fix your attention on the vertical vibrations of a horizontal string of length L and choose coordinates such that the origin is at one end of the string. In that case, y = 0 at x = 0 and at x = L. It is often simplest to solve an equation such as the wave equation by guessing at a reasonable solution and seeing whether it works. A plausible guess that comes from looking at strings in motion is that each particle executes SHM in phase with one another. This means that all the particles are at their maximum positions at the same time. Such a guess would imply that

$$y = f(x)\cos(\omega t)$$

is a possible solution, if f(x) is chosen properly. When this trial solution is differentiated twice with respect to time, an equation is obtained that involves only f(x). The term involving the time can be factored out. The resulting equation is of the form

$$\frac{d^2 f}{dx^2} = -\frac{\omega^2}{c^2} f.$$

This equation is of the same form as the one used for SHM. The only difference is that x is involved instead of t. It is the equation for a simple harmonic oscillator of unit mass, in which the spring constant has been replaced by ω^2/c^2. A solution is chosen in the form of

$$y = A \sin kx,$$

where A is an arbitrary constant and k = ω/c. The choice is made of the sine, rather than cosine, in order to satisfy the boundary condition at x = 0. It is still necessary to satisfy the boundary condition at x = L. Thus, sin(kL) must be 0. This is easy to satisfy because the sine function becomes 0 when

$$kL = \frac{\omega L}{c} = n\pi, \quad \text{where } n = 1, 2, 3, \cdots$$

The solution just obtained for a particular value of n is called a normal mode. It should be realized that any superposition of normal modes is an acceptable solution. The musical quality of a violin string, or that of any stringed instrument, has to do with the presence of a variety of normal modes oscillating together. The frequency of each mode is the angular frequency, ω, divided by 2π. Each normal mode has a frequency that is an integer multiple of the lowest frequency mode, also called the fundamental frequency. The others are called overtones. Since these frequencies are integer multiples of the fundamental frequency, they are said to be harmonically related. They are often called harmonics of the fundamental. This frequency pattern is responsible for the musical quality of stringed instruments. The different sounds produced by different instruments come from the relative strengths of the harmonics. The vibrations of a drumhead are different in that the overtones are not harmonics of the fundamental. Newton's laws help us understand musical theory.

The diagram below shows the shapes associated with the normal modes of vibration of a string. Every mode except the fundamental has stationary points called nodes. These are points in the string's interior for which the amplitude of vibration is 0. If you could grab the string at a single point corresponding to a node, the vibration would be unaffected. The points midway between nodes are called antinodes. These are the points for which the amplitude of vibration is a maximum.

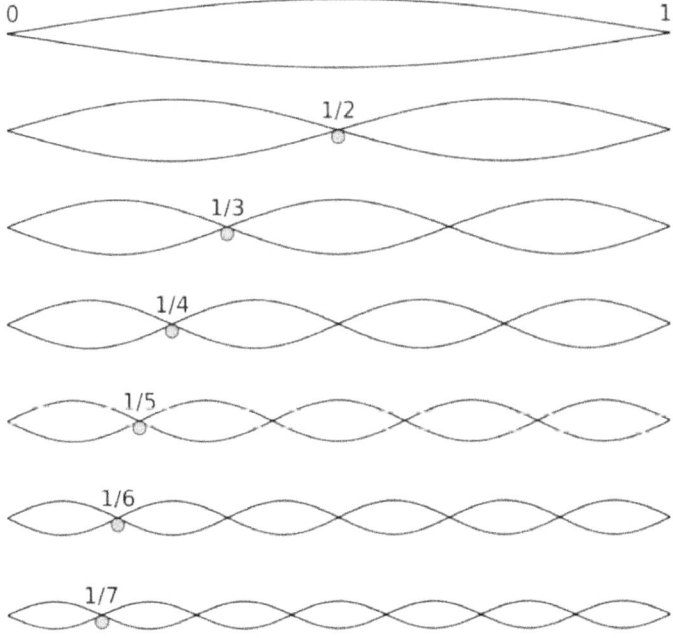

Polarization

A horizontal string can vibrate horizontally as well as vertically. The fact that a single direction, namely, the y direction, is used to characterize a possible motion is for the purpose of simplification. In general, motion along the z direction of a three-dimensional Cartesian system is also possible. Motion in one of these two directions has no effect on the other. Motion along y and motion along z can be superposed. Imagine a violin string that is vibrating vertically in the same normal mode as its horizontal vibration. The motion of a particle on the string will not only depend on the amplitudes of these two vibrations, it will also depend on their relative phases. Consider the following example:

$$y = A \sin kx \cos \omega t, \quad z = B \sin kx \cos \omega t.$$

These equations imply the following: At any instant of time, and at any position along the string, the ratio of y to z is A/B, a constant. In other words, all the particles are moving back and forth along lines that are parallel to one another. It is similar to the motion considered originally in which all the particles were moving up and down along the y direction.

A more interesting possibility is one in which the two motions are 90° out of phase. In this case, a possible motion is given by

$$y = A \sin kx \sin \omega t, \quad z = B \sin kx \cos \omega t.$$

The time can be eliminated from these equations as shown here:

$$\frac{y^2}{A^2 \sin^2 kx} + \frac{z^2}{B^2 \sin^2 kx} = \sin^2 wt + \cos^2 \omega t = 1$$

It can be seen that the particle at the location specified by x moves along an ellipse. The orientation of all these ellipses are alike, in that their major axes are parallel to one another. The size depends only on how close the particle is to a node. It is possible to think of the elliptical motion as the most general case, in that ellipses of any orientation or magnitude are possible. Linear polarization (those that were already considered) and circular polarization are then special cases of these. If you wonder how you can be sure that an ellipse is possible for which the major axis is not along y or z, think about the fact that y and z were picked in arbitrary directions in the first place. These modes can be constructed by superposing linear vibrations, parallel to the y and z axes, that are neither in phase nor 90° out of phase.

These ideas can be applied to traveling waves as well. Traveling waves can be made of any shape. If a traveling wave has a sinusoidal shape, it is also sinusoidal in time. These waves can be linearly polarized, circularly polarized, or elliptically polarized. If the wave is made up of sections, each of which is polarized differently from another, and if it is a sufficiently chaotic arrangement, it can be said that the wave is unpolarized. It should be noted that polarization is a phenomenon associated with transverse waves. Sound waves in air are longitudinal, in that the vibratory motion is along the direction of propagation. A sound wave cannot be polarized.

Many modern cameras make use of filters that are called circular polarizers. You might be interested in how this can be done when the incident light is unpolarized. Imagine that the first layer of the filter through which the light passes is a linear polarizer. It absorbs half of the incident light. The rest of the filter consists of a quarter wave plate. This is a crystalline material that slows up a wave polarized in the x direction more than one polarized in the y direction. If the two waves arrive in phase, they emerge 90° out of phase. The linearly polarized light that enters this part of the filter makes an equal angle with these two special axes in the crystal. Thus, the two waves are in phase and of equal amplitude when they enter the crystal. When they emerge, they are of equal amplitude but 90° out of phase. The result is a circularly polarized wave. It is probably not practical to design a filter that works equally well for all frequencies of light. It may work perfectly only for a limited range of frequencies in the midrange of the spectrum of white light.

Reflections

Consider a string that extends from x = 0 to infinity. The end at x = 0 is fixed in place. If a traveling wave is approaching this end, it will be reflected. In order to prove this statement, use is made of a mathematical trick that avoids the use of a boundary condition. The trick consists of concocting a situation on an infinite string that creates a node at x = 0. Consider the figure below. The wave on the right is a real wave approaching the fixed end of a string at x = 0. This is equivalent to the situation in the diagram of two waves on an infinite string moving as shown. When the two waves reach the origin, their superposition will produce a node there. The wave equation is satisfied, as is the boundary condition for all points on the string. The real situation is equivalent to the unbounded string in all respects. After a short time, only the wave traveling to the right will exist in the string. This is equivalent to a reflection with a reversal in sign. The wave is said to have undergone a 180° phase change upon reflection. The language stems from the fact that a sinusoidal wave reverses sign every 180°.

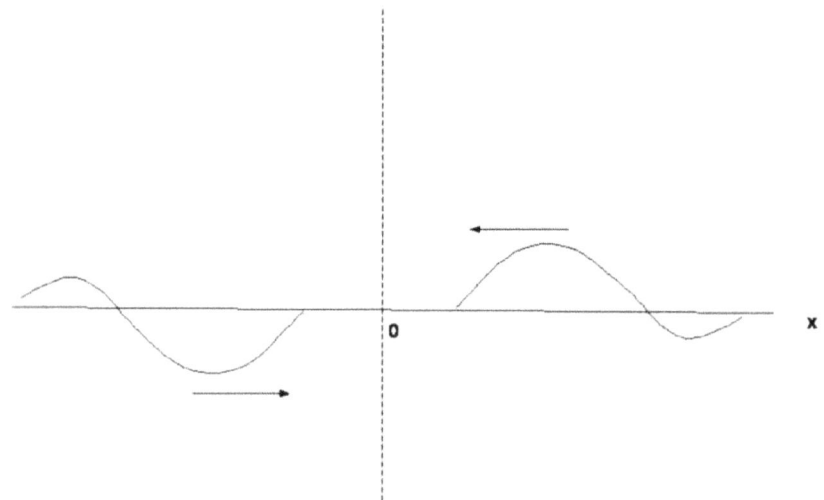

Partial Reflections

Reflections can occur when there is a sudden change in the medium as when a light wave hits a pane of glass. These conditions can be simulated on a string. Imagine a string whose linear density (mass per unit length) changes abruptly. As you have seen, the wave speed depends on the density. Consider a wave reaching a junction where the density suddenly becomes large. In the mathematical limit in which the density becomes infinite, this junction would act like a fixed wall. This would be equivalent to a fixed boundary and result in a 180^0 phase change for a sinusoidal wave. If the density is finite, there is still a reflected wave. In this case, the amplitude of the reflected wave is smaller than that of the incident wave, but there is still the same type of phase change. A portion of the wave is transmitted, and the total amount of energy is conserved. In other words, some of the energy in the incident wave is reflected; some is transmitted.

A similar situation arises if there is a sudden decrease in the density of the string. In this case, the wave speed increases upon transmission. The main difference in the two situations has to do with the phase of the reflected wave. There is no phase change upon reflection when the density of the new medium is lower than the older one. Similar things happen when light gets reflected from glass. There is a phase change in the reflected wave as the light enters glass from the air but none in the reflected ray when the light emerges from the glass back into the air.

Standing Waves

A Graphical Treatment

If you were to pluck a guitar string and wished to know how its shape varied as it vibrated, a normal mode analysis wouldn't be very helpful. A large number of normal modes would be superimposed so that such an analysis would be extremely cumbersome. It is better to break the wave up into two traveling waves and deal with the reflections at the boundaries. Suppose a guitar string is plucked in such a way that at time 0 its shape is known. In other words, at t = 0, y = f(x), where f(x) can be specified as some known shape. If at t = 0 the string is released from rest, it can be considered as the superposition of two traveling waves, as follows:

151

$$y = \frac{f(x-ct)+f(x+ct)}{2}.$$

In other words, the wave is made up of two waves of identical shape that travel in opposite directions. To justify this, first note that the equation is satisfied at t = 0. It is only necessary to show that

$$\frac{\partial y}{\partial t} = 0, \text{ at } t = 0.$$

When the expression for y is differentiated with respect to t, there are two contributions of equal size at t = 0. One of these has a coefficient of negative c, and one of which has a coefficient of positive c. These two terms cancel, which validates the assumption that the speed of each particle is 0 at t = 0.

Suppose, for example, you were to pull a guitar string from the middle until it has the shape OAL, shown below.

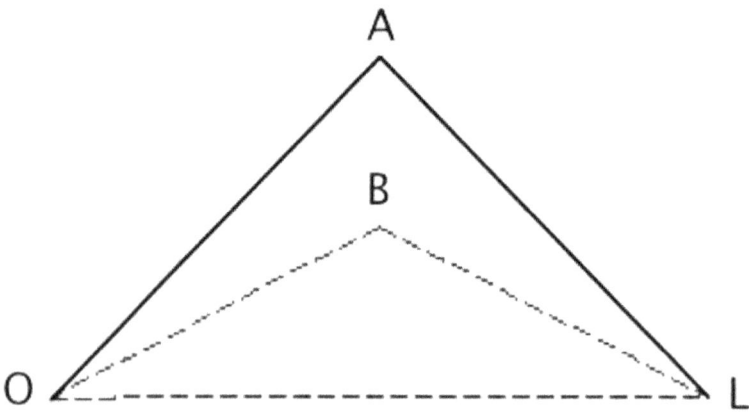

The vertical scale has been exaggerated (tremendously!) for clarity. (The theory is only valid for small amplitude vibrations.) A rigorous treatment also requires that the string doesn't have an abrupt change in slope at the point A. (These are minor details, but are worth noting.) If you pull the string until it has the shape OAL and then let go, the string will satisfy the criteria of the previous paragraph. Its motion can be regarded as arising from the superposition of two traveling waves, each having the shape OBL.

The problem that remains is the analysis of the behavior of the traveling waves and their reflections at the points O and L. The reflections of these traveling waves at the boundaries at O and L are of the form of an inverted OBL. A moment's reflection (a word chosen intentionally) should convince you that these inverted triangles will be followed by triangles identical with OBL. The resulting pattern consists of two waves, of infinite extent, that are in the form of saw teeth, as shown below, to a different scale.

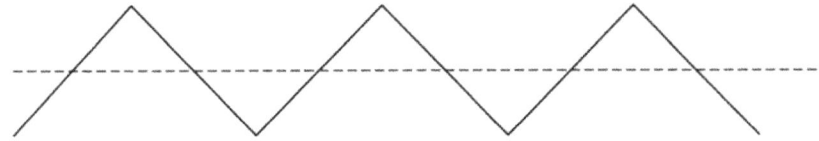

The problem of determining the shape of the string at different times reduces to analyzing the pattern arising from the superposition of two waves of the above shape when they travel in opposite directions. Some thought along the following lines makes the problem simpler: Since the derivative of a sum is the sum of the derivatives, it follows that there are three possibilities for the slope of the string. 1. The slope is the same as that of the line OA. 2. The slope is the same as the line AL. 3. The slope is 0. The analysis is helped by the following considerations. The pattern is symmetrical, and straight strings don't accelerate. This analysis leads to the picture below. The only moving part is the horizontal section. It is traveling with constant velocity.

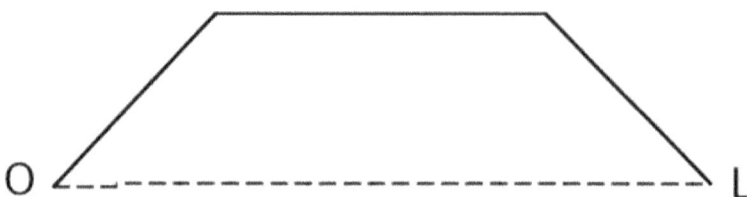

You can check that this picture is consistent with the saw-tooth pattern shown earlier. The string vibrates in the form of a trapezoid that flattens out and then grows again into an inverted triangle. The string becomes flat in the time it takes for the peak of a saw tooth to travel a distance $L/2$ so that the time it takes to make one-fourth of a full vibration is $L/2c$. This is consistent with the period being $2L/c$. You should remind yourself that the period of the fundamental is consistent with this value. Since all the harmonic frequencies are multiples of the fundamental, it follows that every possible vibration of the string will repeat itself in the time $2L/c$.

If you are confused by the fact that parts of the string are moving with 0 speed or with a constant one, and nothing in between, it should be recognized that the discontinuous slope of the string is the problem. It leads to an instantaneous change in velocity, implying a rate of change (i.e., an acceleration) that is infinite. A slight change in the picture, which allows for a continuous slope would have taken care of this difficulty.

There are a few comments about the behavior of this particular vibration that are worth mentioning. The fact that it is plucked at its center makes the pattern symmetrical. If you look at the patterns of the normal modes, it may be seen that every other mode of vibration is symmetrical. The remaining modes are antisymmetrical, which means that they reverse sign when reflected about the midpoint. Thus, the fundamental and the odd harmonics, those whose frequencies are odd multiples of the fundamental, are the symmetric ones. These are the only kinds of vibration that are superposed in a symmetrical vibration. If you want to have even harmonics of the fundamental present in a vibration, it is necessary to pluck the string at some point other than the midpoint.

Chapter 9
The Rotating Earth

The earth has been treated as if it provides us with an inertial frame of reference for the most part. In the treatment of vibratory motion, for example, this can be easily justified. The time scale for a typical single vibration is less than a hundredth of a second. It is very difficult to detect the earth's motion in such a short time. There are phenomena in which the earth's rotation cannot be neglected. The trajectories of missiles and other weaponry are good examples if the time of flight is not extremely short. The weather is a better example of a phenomenon that cannot be understood without taking the earth's rotation into account.

A casual glance at a weather map indicates regions in which the pressure is changing from one place to another. You might ask yourself, "How can this be?" In our everyday experience, the air pressure in a balloon can only be maintained if the air is completely confined. As soon as there is a hole or a tear, the air rushes out and the pressure is reduced to atmospheric pressure. Why, then, is the air in a region of high pressure of the atmosphere not rushing out also? It seems as though the air that descends on us from Arctic regions is held together as in a balloon. It only interacts with warm, moist tropical air at its boundaries. It is at these boundaries, called fronts, that most of our bad weather takes place.

In order to deal with the rotating earth as a reference frame, it is necessary to modify Newton's laws. If a body is at rest with respect to the earth, at the equator, for example, it is accelerating with respect to an inertial frame. There is a net force on it even though it is in equilibrium with respect to the earth. The only way to make Newton's second law seem valid is to introduce an extra force (a fictitious one) called the centrifugal force. There is no body that exerts this force. This is a good clue that the force is fictitious. Another clue has to do with the fact that such a force is proportional to the mass. Although this is an attribute of the gravitational force, Newton didn't regard gravitational forces as fictitious. It will be seen that Einstein had a different view of gravity. Before embarking on a detailed explanation of certain weather phenomena, it is useful to take up a few other matters first.

Scaling

You may wonder, "Why can insects manage to walk with legs that are skinny, while elephants and other large animals need much fatter legs, relative to their sizes?" This type of question brings up the concept of scaling. Consider any object and magnify it in such a way that its shape is preserved. If a linear dimension is doubled, an area is increased by a factor of four (two squared) and a volume by a factor of eight (two cubed). Movies have shown us ants that are about the size of humans. Could such creatures possibly evolve on earth? If they were to increase in size by a factor of 100, their volume, and presumably their weight, would increase by a factor of 1,000,000. The cross-sectional area of their limbs would increase, but only by a factor of 10,000. The pressure on the legs would go up by a factor of 100. It is likely that they would collapse. Evolutionary processes would have to invent new materials or new shapes in order to allow for such an increase in size. This is a scaling idea, one that suggests that certain sizes are optimal for certain shapes.

The scaling idea also suggests why big objects usually fall faster than small ones, and why tiny iron filings practically float in air. The downward gravitational force on such objects scales with the volume, whereas the air exerts a force that scales with the area. Very fine dust takes a very long time to settle, even in still air. Such considerations suggest that birds the size of emus and ostriches have a harder time flying than small songbirds.

You can understand many phenomena by considerations of scaling. Consider, for example, the behavior of coal. It would be very difficult to ignite a lump of coal the size of a chicken egg or larger with an ordinary match. Also, it would be extremely dangerous to light a match in a coal mine that is laden with coal dust. In the latter case, a much greater fraction of the coal is in contact with the oxygen in the air so that once the reaction (carbon plus oxygen into carbon dioxide) is started, it proceeds so rapidly that an explosion is likely. Moreover, the heat from a match flame is dissipated on contact with a large lump of coal. The energy is spread over a large volume so that the lump doesn't get very hot. Yet a speck of dust is likely to get white hot in an instant in the presence of a flame. In fact, the glow of a match flame is due to that very same chemical reaction. If you have a fireplace, you might try the experiment of throwing a piece of steel wool into a fire. It will burn fairly rapidly. A steel poker, however, can be used in the same fire with no apparent reaction. You should try to think of other phenomena in which size is relevant.

Scaling has a bearing on weather phenomena in that clouds consist of water droplets that are so small they can be suspended as though they were almost weightless. In order for rain to form, as in a thunderstorm, it is usually necessary for clouds to grow to a height in which ice crystals are formed. As these grow to a size in which they fall through the cloud, they pick up more water and grow larger. If there are enough updrafts, these ice crystals can move upward and grow into large hailstones before coming down and hitting the ground. They usually melt on the way down and hit the earth as rain.

Pressure and Buoyancy

Every weather station has a barometer, an instrument that measures atmospheric pressure. In order to define pressure, imagine an object that has a flat surface on one of its boundaries. In the absence of any wind, the atmosphere exerts a force on this surface directed inward. This force is proportional to the area. The ratio of the size of this force to the area is a property of the atmosphere and not the surface. This ratio is called the atmospheric pressure and is commonly expressed in pounds per square inch. This force may be attributed to the bombardment of the surface with a huge number of molecules each second. These molecules are moving in random directions and with a distribution of speeds. If you were able to measure the force during a time that is short compared to the time between collisions, the force would be expected to fluctuate violently. The usual definition of pressure assumes a much larger time interval. Most instruments cannot detect any fluctuations at all.

If you were to slice an apple in half, you might think that there could be a net force on it due to the atmospheric pressure. The flat side of the sliced apple has less area than the area of the peel. A moment's reflection should convince you that the vector nature of a force makes the force on both sides the same strength. The component of force in any direction depends only on the cross-sectional area that you see when looking in that direction. A uniform pressure will not yield a resultant force on any real object, no matter what the shape. Nevertheless, atmospheric pressure is responsible for the buoyant force that is responsible for lifting dirigibles and balloons. It isn't the shape that counts. It is the variation of atmospheric pressure with altitude.

In order to understand why the atmospheric pressure depends on altitude, imagine a column of air that extends from the ground to the top of the stratosphere. For the sake of argument, you could imagine that this air column is inside a pipe that is open at both ends. There are two vertical forces on this column of air. One is the gravitational force, namely, the weight of the air column. The other is the force that the ground exerts on it. Since the air column is static (by assumption), the net force is 0. The force on the ground, due to the air column, is just the reaction to one of these forces (which one?). The weight of the air column can be equated to the force on the ground, which is also given by PA. Here, P is the atmospheric pressure at the earth's surface, and A is the cross-sectional area of the air column. If the atmospheric pressure is 14.7 pounds per square inch, then the air in a column 1.0 square inches in area weighs 14.7 pounds.

The same argument can be used for the pressure at any altitude. The pressure depends only on the weight of an air column that extends upward from that altitude. The same reasoning can be used for the way the pressure depends on depth in the ocean. If the approximation is made that the density of ocean water is constant and that g is constant, a formula can be found for the dependence of P with depth, D. The volume of a column of water of length D, and cross-section A, is AD. If the density (mass per unit volume) is ρ, the weight of this column is $DA\rho g$. The pressure contribution arising from this column of water is then $\rho g D$. The atmospheric contribution at sea level, P_0, has to be added to this. The pressure is then given as $P_0 + \rho g D$. It is not possible to develop such a simple formula for the atmosphere. The density of air depends on the temperature and the pressure. To some extent, it depends on its moisture content as well. For these reasons, the atmospheric pressure at a given location can vary by as much as 5 percent in the course of a few days.

The buoyant force on an object immersed in a fluid arises as a result of the variation of pressure with altitude. Imagine a solid object that is completely submerged in a fluid. The pressure forces on the bottom are greater than the pressure forces on the top. The buoyant force is the net force arising from all these contributions. Newton's ideas can be used to show that they are consistent with the conclusions arrived at by Archimedes many years earlier. Archimedes was considering a body that was submerged in a fluid. He might have imagined a second solid body that had exactly the same shape as the submerged one. It only differed in that it had exactly the same density as the fluid. Imagine now that the submerged body were replaced with this imaginary one. Since the new body has the same density as the fluid, it behaves exactly like the fluid itself and remains in equilibrium. This implies that the buoyant force on it is equal to its weight. The statement by Euclid was that the buoyant force equals the weight of the displaced liquid. This is consistent with Newtonian physics in that the buoyant force arises only from the pressure differential on various portions of the sub-merged object. These forces don't depend on the density of the submerged object and are the same for the original object and its replacement.

The same argument can be used for a floating object. In this case, the displaced liquid corresponds to the portion of the object below the surface of the fluid. These ideas lead to a question that often stumps graduate students in physics, as well as some profes-sors. Imagine a Ping-Pong ball that floats on top of a bowl of water inside a bell jar. A bell jar is a kind of vessel that is sealed off in such a way that it can be pressurized by an air pump that is connected to it. To start with, the pressure inside the jar is atmospheric. The pressure is now raised, by pumping air into the container. Assume that neither the Ping-Pong ball nor the water is much changed by the increase in pressure. The question is whether the Ping-Pong ball will be forced deeper into the water by the added pressure or remain the same, or perhaps it will be less deeply submerged? Which do you choose? Do not read the next paragraph if you would like to work this out for yourself.

The wording of the question is what throws most people off. In-stead of thinking of the pressure as pushing on the ball, you should look at the problem as a simple buoyancy question. The Ping-Pong ball is immersed in two fluids, one of which is the water. The other fluid is the air. Both fluids contribute to the buoyancy. People gen-erally recognize the buoyancy of air when dealing with balloons but tend to forget about it when dealing with heavier objects. The contribution to the buoyancy force from the air equals the weight of the air displaced. As the pressure increases, the density of the air goes up so that it becomes more buoyant. The ball would rely less on the buoyancy of the water and rise out of it a bit. It is even con-ceivable that it would leave the water completely and float to the top of the air chamber if the density of the air became greater than the average density of the Ping-Pong ball.

Weather

An advanced treatment of weather phenomena would require the use of sophisticated physical and chemical ideas. Only certain features of air masses and weather fronts will be considered in order to show how Newton's laws come into play in their behavior. The earth's rotation is important in the explanation of this behavior. It is necessary to modify Newton's laws so that they describe motion with respect to a non-inertial frame of reference.

Some weather maps display the barometric pressure in the form of isobars (lines of constant barometric pressure). They provide a two-dimensional view of a three-dimensional system. Weather maps may also show cold and warm fronts that represent the boundaries between masses of air that have had different histories. One mass may come from cold, dry conditions in northern Canada. The other may have spent much time over the warm waters of the Gulf of Mexico. Actually, the boundary between these different air masses is nothing like a vertical wall, which seems to be the implication of the map. It is more like the boundary between sand and water at a beach. The depth of the water increases gradually with distance from the shore. In the case of a front, you may think of the cold air as the sand and the warm air as the water. In the case of a cold front, as the cold air advances, it acts like a wedge that lifts the warm, moist air. As the air rises, it expands and gets colder. The moisture may then condense into clouds.

The question as to why these air masses can preserve their identities instead of rapidly mixing into one another is still unanswered. It will be seen that the inertial forces associated with a rotating frame of reference are responsible for this. This is the subject of the next few sections.

Inertial Forces

The discussion of tidal forces implied that the orbital motion of the earth almost cancels out the effects of the sun and the moon.

The main reason the earth's frame of reference is not inertial is because of its rotation. In other words, tidal forces may be ignored as being less important than the other inertial forces. The earth will be treated as though it formed a frame with constant angular velocity, Ω (capital omega), with respect to an inertial frame.

Consider some vector quantity $\mathbf{A}$, whose components are fixed with respect to the earth. From your point of view, the vector is constant. Its rate of change is 0. From the viewpoint of someone observing from an inertial frame, the vector is changing. You may want to review the section on cross products to verify that its rate of change is

$$\frac{d\mathbf{A}}{dt} = \Omega \times \mathbf{A}.$$

If the letter d is used to denote changes with respect to an inertial frame, a different symbol is needed to denote a change with respect to the earth. The symbol d' (dee prime) will be chosen for this purpose. In the more general case, when the vector A is changing with respect to the earth, the formula becomes

$$\frac{d\mathbf{A}}{dt} = \frac{d'\mathbf{A}}{d't} + \Omega \times \mathbf{A}.$$

This formula applies to any kind of vector quantity. When applied to the position vector, $\mathbf{r}$, it yields an expression for the velocity, namely,

$$\mathbf{v} = \mathbf{v}' + \Omega \times \mathbf{r}.$$

The primed quantity, $\mathbf{v}'$, is the velocity with respect to the earth. Use of the same formula for the vector, $\mathbf{v}$, yields

$$\frac{d\mathbf{v}}{dt} = \frac{d'\mathbf{v}}{d't} + \Omega \times \mathbf{v}$$

$$= \frac{d'(\mathbf{v}' + \Omega \times \mathbf{r})}{d't} + \Omega \times (\mathbf{v}' + \Omega \times \mathbf{r})$$

$$= \frac{d'\mathbf{v}'}{d't} + 2\Omega \times \mathbf{v}' + \Omega \times (\Omega \times \mathbf{r}).$$

When multiplied by the mass of a particle, the LHS of this equation corresponds to the net force on the particle. The first term on the RHS becomes the product of the particle's mass with its acceleration with respect to the earth. It can be seen that this product can be regarded as the net force on it plus two correction terms. One correction term depends only on its position. It is called the centrifugal force. As mentioned earlier, this is clearly a fictitious force that arises because the earth's frame is not inertial. It is a force directed away from the axis of rotation of the earth and is proportional to the distance that a particle is from the axis. It is like the gravitational force in that they both are proportional to the mass and depend only on position. In a sense it combines with the real gravitational force to yield an apparent gravitational force. It is part of the force that yields the measured acceleration due to gravity, if no corrections are made. It is not of much interest insofar as the weather is concerned.

The second fictitious force depends on the velocity of a particle with respect to the earth. It is known as the Coriolis force, F_c, and is given by the formula

$$\mathbf{F}_c = 2m(\mathbf{v}' \times \Omega).$$

This fictitious force plays a crucial role in accounting for the wind. It is perpendicular to the velocity vector so that it has no effect on the speed or the kinetic energy of a particle. Its main effect is to cause a particle to be deflected from a straight path.

The Coriolis Force

The discussion has led to the conclusion that as far as the weather is concerned the earth can be regarded as though it were an inertial frame of reference, except for the fact that every particle has an additional force on it given by the formula

$$\mathbf{F}_c = -2m\Omega \times \mathbf{v} = 2m\mathbf{v} \times \Omega.$$

It is no longer necessary to use primes because there is only one frame of reference being considered. Although the earth's angular velocity is constant, to an excellent approximation, its apparent direction, to an earthbound observer, depends on the latitude. At the North Pole, its direction is vertical. At the equator, it is horizontal.

In general, it points toward Polaris, the North Star. It is the vertical component of the angular velocity that has the largest effect on the wind. If it is assumed that the motion of the wind is horizontal, the effect of the horizontal component of omega is to exert a vertical force on the air. It has no effect on the horizontal motion.

The vertical component of the earth's angular velocity is 0 at the equator. Its magnitude is proportional to the sine of the latitude. It is upward in the northern hemisphere and downward in the southern. to understand how it affects the winds, focus your attention on a region of the planet far from the equator in the northern hemisphere. If a pressure difference arises between one location and another in this region, there will be a horizontal force driving the air from high pressure to low pressure. Molecules will then accelerate down the pressure gradient. As it picks up speed, a molecule will experience a fictitious force causing it to change its course. In the northern hemisphere, the force will be to the right so that the molecule will no longer be moving down the pressure gradient. It will have a component of its velocity vector that is parallel to the isobars. Instead of shooting straight across the isobars from high pressure to low pressure, a molecule will be moving more and more tangentially to these isobars as it picks up speed. If there is no friction, the speed will reach a limiting value, and the motion will be along an isobar. The net force on the particle, from the real and inertial force, will account for the acceleration of the particle as it moves. In other words, if the isobars are straight, the forces will cancel.

It should be appreciated that everything that is claimed here is an approximation, or an idealization. Friction cannot be ignored near the surface of the earth, especially over land where there are mountains, buildings, trees and other obstructions. The wind moves along isobars more closely over the oceans than over land.

In general, the pattern of wind circulation is clockwise around high-pressure regions and counterclockwise around lows in the northern hemisphere. The opposite is true in the southern hemisphere. It is possible to imagine the wind circulating around a low-pressure region in the "wrong" sense. Although the Coriolis force is outward in this case, the pressure forces are in the right direction to provide an inward acceleration. Although this could happen, it is a rare occurrence. The wrong circulation around a high-pressure region isn't possible, since both types of force are in the wrong direction. High-pressure regions can't cross the equator because the Coriolis force there contributes nothing to the circulation.

The analysis has been oversimplified by the assumption that the motion of the air is strictly horizontal. However, it provides a qualitative picture of how air masses can retain their identity by being bounded by a region of circulating air. The mixing of different air masses occurs only at the boundaries of these regions. It is thus possible for a cold dry parcel of air from the Arctic to keep its identity for a long time before it dissipates.

In this section, some of the assumptions that were made in accounting for the winds are examined. For example, are the pressure differences in a given region large enough to account for the wind speeds that are found there? For simplicity, treat a parcel of air as though it can't come apart, as though it were surrounded by a weightless membrane. This is a fiction, but it contains all the meaningful ingredients of the real system. Refer to this system as a parcel of air and imagine it is in the shape of a box, as in the diagram below. Assume the pressure depends on x but is constant in the other horizontal direction.

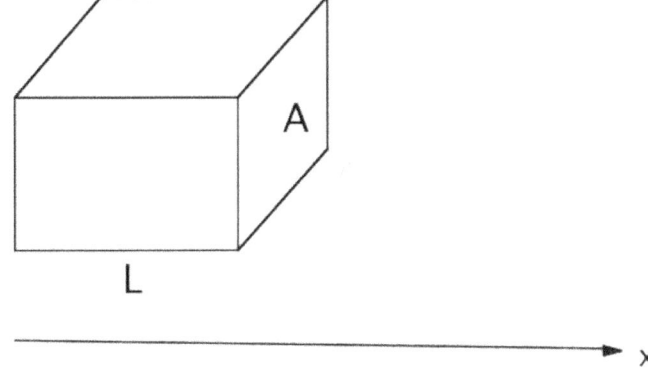

The volume of the box is the product of the area, A, and the length, L. The mass of this parcel of air is then ρAL, where ρ is the density. It is assumed that L and A are sufficiently small that this parcel can be treated as though it were a particle. The pressure gradient causes there to be a net force on it. There will be a force in the negative x direction if the pressure is rising from left to right. The magnitude of the force on any face is the product of the pressure with the area. Thus, the net force to the left equals the product of the area with the increase of pressure in the distance, L. Since L is small, this pressure difference equals its rate of change with distance times the distance, L. The net force in the positive direction due to the pressure gradient is given by

$$F = -\frac{dP}{dx}LA.$$

It will be assumed that the wind has negligible acceleration so that this horizontal force is balanced by the Coriolis force. In order to get the horizontal component of this force, the vertical component of the earth's angular velocity is needed. This is given by the product of the angular velocity with the sine of the latitude, λ (the Greek letter *lambda*). The magnitude of the Coriolis force is then

$$F_C = 2mv\Omega\sin\lambda = 2\rho ALv\Omega\sin\lambda..$$

The assumption that the two forces balance leads to this equation for the wind speed:

$$v = \frac{1}{2\rho\Omega\sin\lambda}\frac{dP}{dx}.$$

The purpose of this calculation is to get a ballpark estimate for the predicted velocity compared to that shown on a weather map. If the calculated value differs from the observed velocity by an order of magnitude (a factor of ten) or more, it is reasonable to assume some other mechanism might be at work.

This formula was checked by consulting a weather map on a windy day. In the test, the wind speed was measured to be 40 mph while the value that was predicted by the formula was 75 mph. This was considered good agreement because the formula is likely to give an overestimate of the speed. It assumes that the air has reached a steady condition in which it is traveling parallel to an isobar. It should also be stated that it also assumes that the isobars are straight, so there is no acceleration if the speed is constant. The main reason that it is an overestimate is that it ignores friction and the possibility that the air is not flowing along an isobar. Even if the Coriolis force and the pressure force were to balance, frictional forces would slow the air so as to reduce the Coriolis force. Evidence for this would be in the fact that there would then be some flow of air from high pressure to low pressure. Look at a weather map in which wind directions as well as isobars are shown. Over the oceans, where friction is small, the flow of air is nearly parallel to the isobars. Over land, there is a component of velocity perpendicular to the isobars. It is greatest where the terrain is hilly and tall building or forests obstruct the flow.

There is still one question that should be answered in connection with the speed of the wind. It has been assumed that the energy gained by the air is provided by the pressure gradient. Can the work done by the pressure force provide the kinetic energy associated with a strong wind? We are only interested in whether some movement from one region in an air mass to another can provide the type of wind speeds that are encountered.

The same parcel of air that was shown in the previous diagram is being considered. The Coriolis force is now being neglected, as it has no effect on the energy of this system. Newton's second law leads to

$$-\frac{dP}{dx}LA = ma = \rho LAa,$$

so that

$$a = -\frac{1}{\rho}\frac{dP}{dx}.$$

The acceleration has been calculated, rather than the force, as it doesn't depend on the length of the assumed parcel of air. Look at the equation below, in which it is assumed that all vectors are in the x direction. The triangular symbol is capital delta. It is used to denote the change in a quantity. The first equality expresses Newton's second law. Note that Fdx is the work in moving along x a distance dx. The next equation equates the work to the change in kinetic energy. The next equation expresses the kinetic energy in terms of the speed. The next equation makes use of the expression for the acceleration in the very first term. The last term involves the change in pressure over the distance dx.

$$adx = Fdx / m = \Delta K / m = \Delta(v^2 / 2) = -\frac{1}{\rho}\frac{dp}{dx}dx = -\frac{\Delta P}{\rho}.$$

The density of the air has been treated as a constant. It is an approximation. Assuming the air starts off from rest, the estimated speed picked up by a parcel of air that starts off with zero speed is

$$v = \sqrt{\frac{-2\Delta P}{\rho}}.$$

Note that the minus sign is necessary since ΔP is negative also. The average density of air at sea level is 1.28 kg per cubic meter. Consider a drop in pressure about 1/100th of atmospheric pressure—thus, $-\Delta P$ = 1.01 x 10 ^{3} Newtons per square meter. This gives a speed of 39.7 meters per second, or about 89 mph. If the pressure change is reduced by a factor of 100, the speed is reduced by a factor of 10. The atmosphere seems capable of generating substantial wind speeds by a movement of air through very small pressure drops.

Commentary

Although much space has been devoted to Newton, his theories, and their applications, it should be realized that this is only a tiny, and elementary, sampling of his work. This is not to imply that Newton carried out all these calculations by himself. Newton opened a door, and other scientists and engineers walked in. The fields of acoustics, hydrodynamics, and mechanics owe a huge debt to Newton. The space program also is almost completely indebted to his work. Mathematicians use both differential and integral calculus for all sorts of problems. In fact, it is hard to think of any branch of the physical sciences for which Newton's work does not play an important part.

Although Newton's theories accounted for many natural phenomena, very little was known about any action-at-a-distance forces other than gravity. Electricity and magnetism didn't seem to play much of a role in any major phenomenon. It wasn't even known that lightning was an electrical discharge. It also wasn't known that electricity and magnetism were related. The field of electromagnetism grew up in the eighteenth and nineteenth centuries. There are many different people who were involved in the early developments, and there are a number of physical laws named after them . However, it took the efforts of one man to combine these different ideas and, by doing that, to make a complete classical theory out of these distinct laws. The man who did that was James Clerk Maxwell (1831–1879), a British physicist whose work had a profound influence on what came later.

Part 2
Maxwell

Chapter 10
Electromagnetic Theory
Light: Wave or Particle

In the seventeenth century, there were two conflicting theories concerning the nature of light. The theory that had the most adherents was the particle theory put forth by Isaac Newton. Newton was aware of Snell's law of refraction in optics. This is a law that describes how a beam of light is bent when it travels from one medium to another, such as from air into glass. Newton was able to account for this bending by assuming that the particles of which the light beam was composed were attracted to the glass by a strong force as the beam got very close to the glass. This would imply that these particles were speeding up as they entered the glass. Newton had already done experiments that showed that white light was made up of all the colors of the rainbow. He assumed different types of particle were associated with the different colors.

An alternative theory had already been proposed by Christian Huygens (1629–1695), a Dutch astronomer, mathematician, and physicist who was quite renowned in all these fields. Actually, Huygens made many contributions to science but is known today principally for his wave theory of light. Huygens's theory was published in 1678, just a few years before the publication of Newton's theory. Huygens was able to account for the different colors by assuming each color was associated with a particular frequency of oscillation.

Newton thought the wave theory couldn't possibly be right. After all, a wave needs a medium, like water or air or a string. Light propagated through empty space. What could be oscillating? Perhaps most other scientists agreed with Newton or were swayed by his enormous prestige. There were other reasons to doubt the wave theory. If light were a wave, shouldn't it be possible to super-pose two light beams in such a way that there would be dark regions coming from destructive interference? This would arise from the superposition of a crest of one wave with that of a trough of the other. Nothing like that had been observed.

The two competing theories were both able to account for the way that light bends as it passes through an air-glass interface. The formula for the amount of bending had been discovered about five years before Newton's birth. The phenomenon is called refraction, and the formula, known as Snell's law, can be expressed in terms of the angles in the figure below: The incoming ray is usually referred to as the incident ray, and the outgoing one is called the refracted ray.

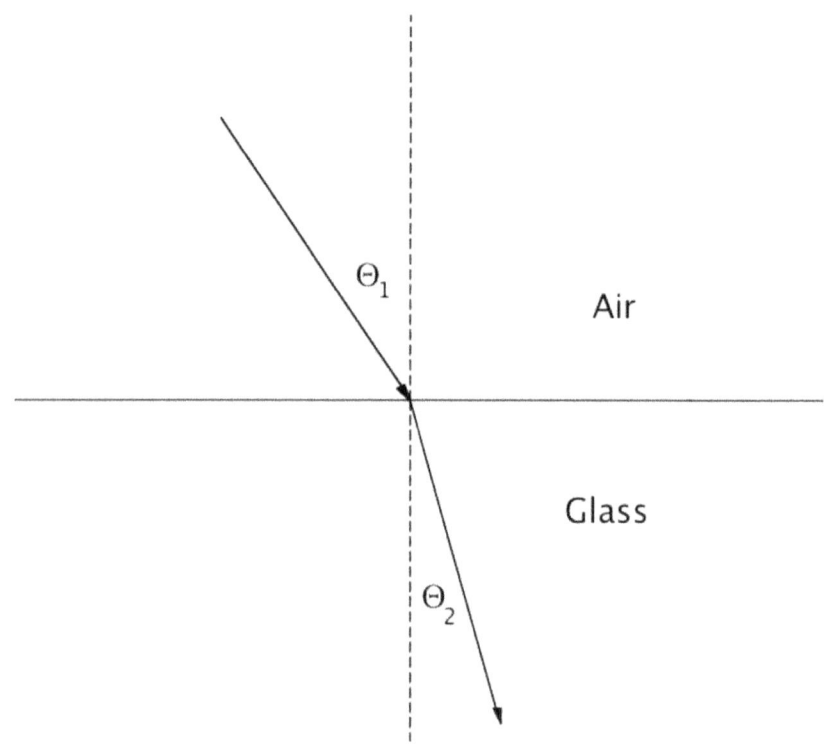

Snell's law can be written as

$$\frac{\sin \theta_1}{\sin \theta_2} = \frac{\eta_G}{\eta_A},$$

where η (the Greek letter *eta*) is a constant for a given medium. It is called the index of refraction. The index of refraction of a vacuum is exactly 1. The index of refraction of air is so close to 1 that the RHS is usually not written as a fraction.

Newton's explanation of the refraction is considered first. Assume that the upper arrow in the diagram represents the velocity vector of an incoming particle. The lower arrow has the proper direction but the wrong magnitude, if it is to correspond to the velocity vector of the particle in the glass. It must be made larger such that its horizontal component equals that of the incident particle. This follows from the assumption that there is no horizontal force acting on the particle. Let v_A and v_G represent the speed of the particles in air and glass, respectively. Let v_H be the horizontal component of each velocity vector. Then Snell's law takes the form

$$\frac{v_H / v_A}{v_H / v_G} = \frac{v_G}{v_A} = \frac{\eta_G}{\eta_A}.$$

Thus, Newton predicted that the index of refraction of a material was directly proportional to the speed of the light particles in that medium.

The wave theory of light predicts that light waves travel faster in air than they do in glass. It is similar to the behavior of waves on a string: the higher the density the slower the speed. Huygens's theory of refraction is illustrated in the diagram below:

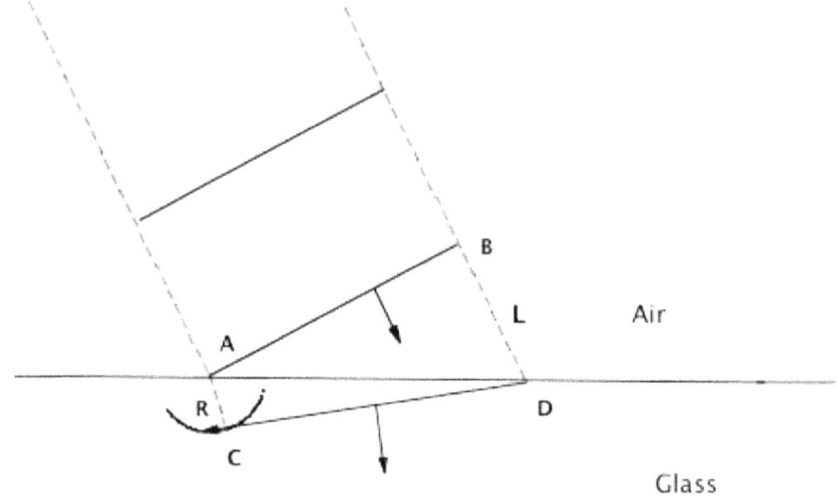

The diagram depicts a ray of light bounded by the dotted lines. The solid lines denote wave crests of a sinusoidal wave. The upper arrow shows the direction of propagation of the wave in air. The crest AB has just reached the glass-air interface at the point A. The wave travels much more slowly in glass than it does in air. In the time it takes for the crest at B to reach the point D, a distance denoted by L, the crest at A can only travel a distance R. The ratio of R to L is in the ratio of the wave speed in glass to the speed in air. The crest at A can only reach a point on the arc shown of radius R. The crest at that time must lie along a line that extends from point D to somewhere on the arc. The arrow that denotes the direction of propagation must be closer to the vertical than the one in air. The wave theory of Huygens causes bending toward the normal (perpendicular) when a wave slows up. An experiment that determines the speed of light in glass could decide the issue.

Huygens's ideas are consistent with Snell's law. In order to show this it is first necessary to show that the line DC makes contact with the arc tangentially. In that case the angle ACD is a right angle. The tangential contact is the only one that makes sense, as can be recognized if the glass were replaced by air, in which case R = L, and there is no refraction. Note that angle BAD is Θ_1. This can be seen from the fact that its sum with angle ADB is 90^0. By the same reasoning you should be able to prove that angle ADC is Θ_2. This leads us to the following equations

$$\frac{\sin \theta_1}{\sin \theta_2} = \frac{L/AD}{R/AD} = \frac{L}{R} = \frac{v_A}{v_G}.$$

Use of Snell's law leads to

$$\frac{v_A}{v_G} = \frac{\eta_G}{\eta_A}.$$

Huygens's theory thus implies that the speed of light in a given medium is inversely proportional to the index of refraction. For a long time scientists had no way of deciding between the two theories. Newton's theory just seemed more plausible to most of them. They had a rude shock early in the nineteenth century when Thomas Young (1773–1829) demonstrated the wave nature of light. His famous double slit experiment could only be explained in terms of Huygens's wave theory. You are encouraged to read about Young's double-slit experiment on the Internet.

Young's experiment created an uproar among scientists. It suggested that a vacuum was not nothing. There must be a medium, they thought. Although this medium had no measurable mass, the conclusion was that it existed and efforts were made to discover its motion with respect to the earth. There was no idea in the early nineteenth century that light had anything to do with electricity and magnetism. Research in this area was in its infancy. It is the connection between these different phenomena that will be the subject of the next section. The coverage of the topic of electricity and magnetism will be superficial. The emphasis will be on those subjects that led to the theory of light waves. Those topics that have little bearing on these developments will be skipped.

Electromagnetism

The earliest electrical experiments date back at least to the time of the ancient Greeks. These experiments were connected with the behavior of amber after it had been rubbed. In fact, the word for amber in Greek is *elektron*. If a piece of amber is rubbed by fur, it is capable of attracting bits of paper and other light materials. Rubbing a glass rod with silk produces a similar affect. A thin stream of water from a faucet can be made to swerve if such a rod is placed close by. Today we refer to these phenomena as electrostatic and account for them by the exchange of charged particles, electrons, from one material to another. Almost everyone is familiar with the atomic nature of matter and realizes that all atoms have electrons. Atoms are electrically neutral because they contain an equal number of electrons and protons. Protons and electrons have opposite charges that are equal in magnitude. The net charge is, therefore, 0. Although there are strong electrical forces holding atoms together, their macroscopic behavior is as though they were uncharged.

Like charges repel one another while opposite charges attract. A common electrostatic phenomenon is the shock that you might get after walking across a carpet and then touching a doorknob. A spark often accompanies this exchange of charge.

Benjamin Franklin (1706–1790) was the first to recognize that there are two types of electrical charge. One type he called positive, the other negative. His choice resulted in a negative charge for the electron. Franklin is also credited with the discovery that lightning is an electrical phenomenon. It was Charles Coulomb (1736–1806) who discovered that charged particles obey a force law similar to the one involved in gravitation, in that the size of the force between charged particles is inversely proportional to the square of the distance between them. The force is proportional to the product of the net charge on each particle. Coulomb's law can be put in the form

$$F = k \frac{q_1 q_2}{r^2},$$

where q is used to denote charge, and k is a constant of proportionality. Note that the force is one of repulsion if the product of the charges is positive. In all other respects, Coulomb's law resembles Newton's law of gravitation. The constant, k, depends on the system of units. In the MKS system of units, the numerical value of this constant is about 9×10^9. In this system of units, the force law is written as

$$F = \frac{q_1 q_2}{4\pi\varepsilon_0 r^2},$$

where ε_0 (epsilon-subzero) is a constant that needs to be determined experimentally. The strange appearance of this equation is a source of confusion to the beginner. Why write a constant in such a strange way? Why wasn't the unit of charge chosen to make the constant of proportionality equal to the number 1? Hopefully, the reasons will become clearer shortly. The unit of charge in the MKS system is called the coulomb. Its definition involves an experiment involving electric currents, a topic that will be discussed later.

Compared to the gravitational forces between electrons and protons, the electric forces are huge. In order to appreciate this, imagine it were possible to separate all the electrons from all the protons in a gram of hydrogen. The electrons are placed into one container and the protons into another, about one meter from the first. It will be seen that this is an almost impossible task. There would be about 6×10^{23} atoms in each container. The charge on the proton is roughly 1.6×10^{-19} coulombs. This results in a charge of about 10^5 coulombs in each container. The force on each container would be roughly 9×10^{19} Newtons . This is about 2×10^{19} pounds. Not only is this force unbelievably large, it is almost impossible to imagine any container that wouldn't be blown apart by the pressure caused by the repulsion of the charged particles inside. Even if the two containers were placed 100 km apart, the force of attraction would be about 2 billion pounds, or 1 million tons. These figures are given to make it clear that when electrical forces are involved with atomic particles, it is a good approximation to neglect the gravitational force between them. It should also be clear that only a very small fraction of the atoms are missing electrons when dealing with positively charged bodies.

The electrical behavior of metals is very different from that of materials like glass or porcelain. Electrical charges can move readily in metals. Static charges on metals can quickly redistribute themselves. Loosely speaking, materials can be divided into two categories, conductors and insulators. There are some materials, such as silicon and germanium, called semiconductors, which conduct electricity at high temperatures but are insulators at low temperature. The type and concentration of impurities in these materials play a vital role in their electrical properties.

The Electric Field

Two action-at-a-distance forces have been encountered. The first is that of gravitation, and the second is called electrostatics. Today, most physicists have a problem with the notion of action-at-a-distance. Instead of a direct interaction, they prefer to think of some agent, known as a field, as carrying out this interaction. In a sense, the interaction consists of two parts. The first part consists of one body creating a field. The second part consists of the field acting on other bodies, possibly the body that set up the field itself. Although the field concept seems artificial, it allows for the possibility of a delayed interaction. Newton didn't allow for this possibility when he postulated the third law (action and reaction), but today most scientists accept the idea. If the sun were suddenly to explode into two pieces, it is generally believed that distant planets wouldn't respond to this changed condition immediately. It is believed that some sort of gravitational wave would occur that would propagate outward and affect the planets at a later time. In discussing the gravitational field arising from the sun, it can be said that the force field is a vector quantity that depends on position, and maybe on time. For the stationary sun, the time dependence is absent, and the field vector is pointing toward the sun, its magnitude depending only on distance. To find the force that this field exerts on a particle, it is only necessary to multiply this field by the mass of the particle. In other words, the field has the units of acceleration. In fact, in the case of gravity, the field vector is just the quantity that has been called **g**, the acceleration due to gravity. The only difference is that the letter g is usually applied to the field of the earth, not the sun.

The case of an electric field is very similar to that of the gravitational field. The difference is that the electric field is created by a charged body, and it has the units of force per unit charge. The force on a particle in a given field is obtained by multiplying its charge by the vector field. If the charge is negative, the force on the particle is in a direction opposite to that of the field. The electric field is denoted by the letter **E**. In general, it is a function of position and time, but in electrostatics, the time dependence is absent. The idea of an electric field was not commonplace in the early days of electrostatics. In Coulomb's time, the force was considered a direct interaction.

Although the electric field has the units of a force-per-unit charge, the field cannot always be tested directly by merely measuring the force produced on charges that are brought into a given region. In order to understand this, assume that you want to measure the field produced by some charge distribution in the vicinity of an electrical conductor. If you were to put a charged particle near the conductor in order to determine the field, the presence of the charge would cause the charges on the conductor to redistribute. The field that was to be measured would be disturbed by the test charge that was used to measure it. You might think of using a smaller charge in order to cause less of a disturbance, but there is no way of using a limiting procedure, since the minimum charge available is that of a proton or electron.

It is worth mentioning here that there are certain misunderstandings concerning the nature of physical theories. It is often assumed that physics deals only with quantities that can be measured. This is not always the case. It is probably not possible to measure the electric field inside a light wave, for example. However, the assumption that these fields exist enables a scientist to predict many properties of light and how light interacts with matter. The success of a theory has to do with the variety of phenomena that it accurately predicts. If you hear someone begin an argument with the statement "Science proves...," you can be confident it is not coming from a scientist. Science may prove that some theory doesn't fit experimental evidence. It never proves some theory is the final word.

Once physicists accepted the field concept, they realized that Newton's third law had exceptions. Since forces didn't propagate instantaneously, it was necessary to modify the notion of momentum conservation. It was possible for some particles to change their momentum without a compensating change in others. However, it turned out that the law of momentum conservation was rescued. It was discovered that the field itself could have momentum. It usually isn't noticed at the macroscopic level. For example, sunlight carries momentum and exerts a force on a body when it is absorbed or reflected. The effect is so small that delicate instruments are needed in order to measure it. However, it is important when analyzing the behavior of electrons and atomic particles.

Gauss's Law

In Chapter 3 it was shown that the gravitational field of an object with the symmetry of a sphere produces the same type of gravitational field as that of a particle. The proof was based on an analogy between the gravitational field produced by a particle and the velocity pattern in a fluid produced by a "fluid eater." This proof was put forth by the great mathematician Carl Friedrich Gauss (1777–1855). Because of the similarity between Newton's law of gravitation and Coulomb's law, the argument used by Gauss can be applied to electric fields. Since electric charges are of two types, it is necessary to think of the analogy in the following way: Positive charges are analogous to sources of fluid. They produce an electric field that is directed away from the charge. Negative charges act like sinks or destroyers of fluid. They produce electric fields that are directed inward.

In order to give a precise statement of the law put forth by Gauss, it is first necessary to define a Gaussian surface. For this purpose, it is helpful to think of a rubber balloon or a tire tube. Such a surface forms a boundary that divides space into two regions, an inside and an outside. A Gaussian surface is not real. It exists in our imagination. If there is a net positive charge within a Gaussian surface, there is a net outward "flux" of the electric field through this surface. It is similar to the fact that if a fluid is being produced in a region that is already filled with fluid, there must be a net outward flow through the boundary.

Gauss's law translates the statements in the last paragraph into an equation. The electric field vector is analogous to the velocity vector of a fluid. The rate of flow requires a bit of discussion. For a fluid in a pipe, it is only necessary to multiply the speed times the cross-sectional area to obtain the flow in units of cubic feet per second, for example. If the velocity vector at the Gaussian surface is not perpendicular to the surface, it is necessary to use the perpendicular component of the velocity before multiplying by the area. In addition, if the velocity vector varies from point to point on the Gaussian surface, it is necessary to break up the area into sections and sum the contributions from each section. The mathematical idea behind this kind of sum is an integral. Be assured that higher math will not be necessary for an understanding of this concept.

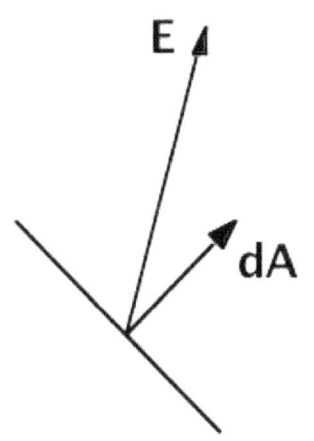

It is helpful to represent an infinitesimal area, dA, of the Gaussian surface by a vector. If the piece is small enough, the surface can be considered flat. The vector dA in the diagram is pointing in the outward direction with respect to the Gaussian surface. The quantity that is analogous to the rate of outward flow is called the electric flux. The electric flux through this portion of the surface is then

$$\mathbf{E} \cdot \mathbf{dA}.$$

The outward flux through the entire Gaussian surface can be written as an integral, as follows:

$$\text{Outward flux } = \oint \mathbf{E} \cdot \mathbf{dA}.$$

The circle through the integral sign is a reminder that the contributions from the entire Gaussian surface should be summed.

This outward flow is equal to the net production rate of all the sources within this surface. If there are sinks, their consumption rate has to be subtracted. Since a positive charge is considered to act as a source, the net outward flux is proportional to the net positive charge that is enclosed. The constant of proportionality depends on the choice of units. In the MKS system of units, Gauss's law is written as follows:

$$\oint \mathbf{E} \cdot \mathbf{dA} = \frac{q}{\varepsilon_0}.$$

The denominator of the fraction is read as epsilon-subzero, or epsilon-naught. In this system, the charge is expressed in coulombs, and the field is in newtons per coulomb.

It is a simple matter to show that Gauss's law is consistent with Coulomb's law. Symmetry arguments may be used to show that the field due to a point charge only depends on the distance from it and that it points directly away from the charge (if positive). If a Gaussian surface is chosen in the form of a sphere of radius r, with the charge at the center, the integral is trivial. It is just the strength of the field times the area of the spherical surface. Gauss's law leads to

$$E4\pi r^2 = \frac{q}{\varepsilon_0}, \quad E = \frac{q}{4\pi\varepsilon_0 r^2}.$$

This last equation is equivalent to Coulomb's law in MKS units. You can now understand the reason for the strange appearance of Coulomb's law in these units. If the factor, 4π, was not included in Coulomb's law, it would show up in Gauss's law. Gauss's law turns out to be more important.

In an earlier chapter, use was made of the fact that the gravitational force was conservative. Since electrostatic fields are also conservative, it is not necessary to use symmetry arguments when applying Gauss's law. This point can be illustrated by a contradiction. It will be shown that if the strength of the field from a point charge is stronger in some directions than others, that the force field will not be conservative. See the diagram below.

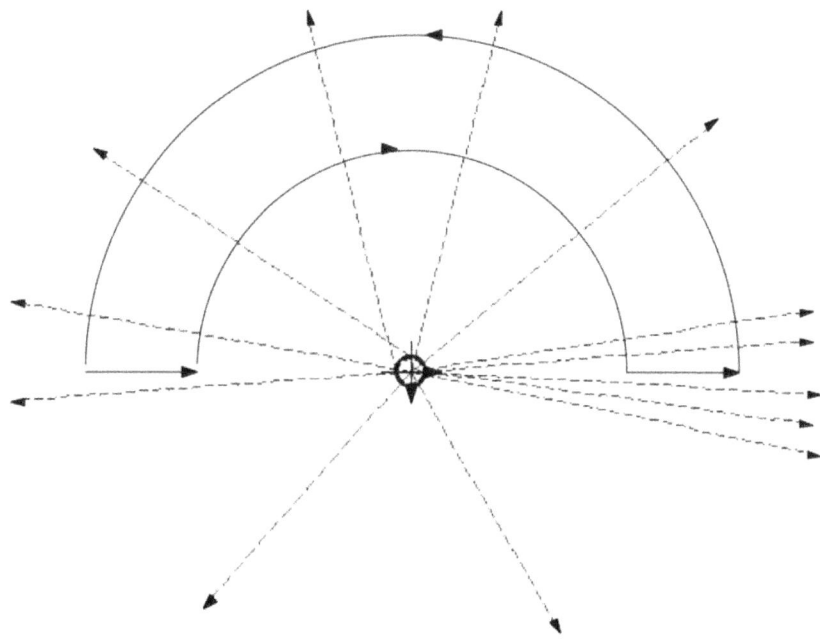

The dotted lines are in the direction of the assumed electric field vectors arising from the point charge at the center of the diagram. The field is strongest where the lines are closest together. It is easy to prove that this force field couldn't be conservative. To see this, imagine a positively charged particle is taken around the path shown by the solid lines in the drawing. No work is done on this particle along the curved portions of the path. This comes from the fact that the electric force is perpendicular to the direction of motion. The only contributions to the work come from the two straight segments. For the given field, the particle would be given more energy at the right end than it gives back at the left end. Such a force field is not conservative. You should conclude that the force must be of equal strength in all directions in order to be conservative.

Historically, Coulomb's law preceded Gauss's law. However, the great physicist James Clerk Maxwell (1831–1879) took the point of view that Gauss's law is more fundamental, in that it seemed to be valid for all types of electric field. Coulomb's law could be derived from Gauss's law and the fact that electrostatic fields are conservative. Gauss's law has interesting consequences. For example, it can be shown that if a net charge is placed on a bar of copper, charges will move around in such a way that no charge resides in the interior. All the charge will be confined to the surface. The argument goes as follows: Since charges are free to move in a conductor, there are no fields inside such a conductor in a static situation. Current flows in a conductor whenever a field is present. Since nothing is flowing in a static situation, there is no field in the interior of the metal. If a Gaussian surface were constructed lying entirely inside the conductor, there would be no flux through this surface. Therefore, there is no charge within it. It is easily concluded that there can't be any charge in the interior of the metal. Strictly speaking, this reasoning applies to a homogeneous metal. If a bar is used that contains a junction between two different metals, it is possible that there is an electric field at the junction. Chemical forces at the junction can balance the forces arising from such a field.

Electric Potential

Consider the gravitational field produced by the sun. The potential energy of a planet depends only on its location in this field and its mass. In fact, the potential energy is directly proportional to the mass. It could be said that the potential energy of any object placed at a particular point in the solar system is a product of the body's mass with a quantity that only depends on its location. This quantity is called the gravitational potential. In general, all the matter in the solar system contributes to the gravitational field. The gravitational potential at a given point in the solar system just depends on the gravitational field produced by these bodies, not on the object that is placed there.

In electrostatics, a similar quantity can be defined, the electric potential, denoted by the symbol V. It is the potential energy-per-unit charge. The reason for the symbol V is that its MKS unit is the volt, a unit that means joules per coulomb. Thus, if a particle of charge 5.0 x 10^{-6} coulombs is placed at a location where the electrical potential is 12 volts, its potential energy is 60 x 10^{-6} joules. Note that only differences in potential really matter because the zero of potential can be chosen to be located wherever it is convenient. The zero of potential arising from a point charge is usually taken to be at an infinite distance from the charge. For the case of a charge, Q, the potential can be written as

$$V = \frac{Q}{4\pi\varepsilon_0 r}.$$

In this equation, r is the distance from the charged particle. If the field is created by several point charges, you can take advantage of the fact that the potential is a scalar. The contributions coming from different charges are added to one another to get the potential arising from the whole system.

If the magnitude and direction of an electric field, **E**, is known everywhere within a given region, the difference in potential between any two points in that region can be obtained from the formula

$$V_A - V_B = \int_A^B \mathbf{E} \cdot d\mathbf{r}.$$

To interpret this formula, imagine a path that connects point A with point B. The path is broken into infinitesimal steps. The letter *r* denotes the position vector; d**r** denotes the change in position vector along a step. Thus, each term in the integral is the work done on a unit charge in a given step. Note that this integral doesn't depend on the path chosen because the electrostatic field is conservative. This implies

$$\oint \mathbf{E} \cdot \mathbf{dr} = 0.$$

The circle on the integral sign means that the integral is carried around a closed path, a kind of loop, in space. If this integral were not zero for an arbitrary loop in space, the field would not be conservative and a potential wouldn't have any meaning. If this point is not clear, consider a loop that consists of one path from point A to point B and another path from point B to point A. If these contributions didn't cancel, the voltage difference between these two points would have no meaning.

Is the electric field produced by a car's storage battery conservative? The potential difference between its terminals is claimed to be 12 volts. This suggests that the field must be conservative. If there is no current flowing between the two terminals of the battery, the field is electrostatic, in that all the charges involved are stationary. As far as the region outside the battery is concerned, the field is produced by these stationary charges. The answer is yes, as long as only the region outside the battery is being considered. In this way, the question of what is happening chemically within the battery is avoided.

There is a difference, however, between the field produced by a battery from one that is purely electrostatic. It is not possible to measure the potential difference between any two points in an electrostatic field with an ordinary voltmeter. The presence of the meter disturbs the field that already exists by redistributing charges on its surface. These voltmeters rely on some current flow, and there is none once the field becomes static. When a battery is present, the potential difference between its terminals is maintained, no matter what happens in the external circuit, as long as the battery isn't being drained by a strong current flow. If a voltmeter is connected to its terminals, the situation is no longer static, as charges will be flowing through the voltmeter.

The potential difference between any two points on a conductor is zero, if no current is flowing. This follows from the fact that there is no field inside a conductor under static conditions. If a room is surrounded by a conducting material, the objects inside are sheltered from strong electric fields arising from charges outside the room. In actual practice, it is not necessary to surround the room with a solid conductor. A conducting cage is usually sufficient.

It has been shown how to determine the potential once the field is known. Actually, it is usually simpler to determine the field once the potential is known. The potential is a scalar, and scalars can be added more easily than vectors. If a charge distribution is known, the potential can be determined by adding the contributions from each charged particle. The fields can then be obtained by differentiation. For example:

$$E_x = -\frac{\partial V}{\partial x}.$$

This is a partial derivative because the other coordinates are being treated as constants.

The Magnetic Field

The history of magnetism is quite complicated. Although magnetism was encountered over three thousand years ago, very little was understood about it until the beginning of the nineteenth century. The earliest experiences with magnetism were with small permanent magnets that were discovered on, or near, the surface of the earth. These crystalline materials are known as lodestones. They are associated with a material called hematite. It has been suggested that lightning strikes are responsible for magnetizing these crystals. The earth's magnetic field is much too weak to magnetize them. Another possibility is that they were magnetized in some other region of the solar system and were borne here on meteors. Lodestones were known for their ability to attract iron. When the lodestone was shaped into a bar, or needle, it was discovered that the bar would point in one particular direction. This led to its application as a compass needle.

A bar magnet behaves in a way that is similar to a rod with opposite electric charges at each end. The attribute that corresponds to an electric charge is called a pole. The end of the magnet that pointed roughly northward was originally called the north-seeking magnetic pole. Now it is simply called the north pole. North and south magnetic poles are analogous to positive and negative charges. Like poles repel one another, and unlike poles attract. A rod with opposite charges at its ends will feel a torque tending to line it up with a uniform electric field—in the same way a compass needle lines up with a magnetic field. The problem with this picture is that no one can find a particle with only one pole. If a bar magnet is cut in two, each new end seems to have a pole. Magnetic monopoles have been sought for more than a century. Some scientists have claimed that they have evidence for them, but their observations haven't been corroborated. It wasn't until after the invention of the electric battery, near the beginning of the nineteenth century, that scientists gained a deeper understanding of magnetic phenomena.

Magnetic fields were once measured in terms of the forces acting on these magnetic poles. For example, a long needle-like bar magnet could be inserted into a region where a field might exist. If it were long enough so that only one end was in the given region, it could be assumed that the force on this end was proportional to the magnetic field. Alternatively, a small compass needle could be used. If the needle were free to turn in any direction, it would line up with the field. The torque on it would be proportional to the strength of the field. These procedures had their problems, and they are no longer used. Physicists now make use of another idea to define the strength of a magnetic field. It is now known that electrically charged particles feel forces when moving through a magnetic field. Recall that, in principle, an electric field can be measured by placing a charged particle in some location and measuring the force on it. Imagine a similar situation in which the charge is allowed to move. The particle will feel an additional force if there is a magnetic field present. It is necessary to consider a series of experiments repeated under identical conditions in which the test charge is given a number of different velocities as it passes a given point. Under these conditions, the force on the particle is given by the following equation:

$$\mathbf{F} = q(\mathbf{E} + \mathbf{v} \times \mathbf{B}).$$

It is named the Lorentz force law after its discoverer, Hendrik Lorentz (1863– 1928), a famous Dutch Nobel laureate. The law can be used to measure both the electric field, $\mathbf{E}$, and the magnetic field $\mathbf{B}$. It should be mentioned that most physicists call $\mathbf{B}$ the magnetic induction, and use the letter H to denote the field. These two quantities differ only by a constant factor in a vacuum. There is no need here for the quantity H , and it will be ignored. The Lorentz force law was unavailable to scientists working in the middle of the nineteenth century. Maxwell was unaware of it. You are being given the benefit of ideas that came later. It would be more difficult for you to understand the theory if a historical approach had been taken here.

Although Newton's laws are assumed to be valid in all inertial frames of reference, it wasn't clear whether electric or magnetic phenomena depended on the motion of the frame. The force on a charged particle in a magnetic field depends on its velocity. It could be moving with respect to the earth and still be stationary with respect to another reference frame. Does that mean the laws are different on this other reference frame? Another alternative is that the laws are valid on all inertial frames of reference but that the fields are different. For example, suppose that there is a magnetic field in a given region but no electric field, as far as an observer on the earth is concerned. A moving charge would feel a force in a direction perpendicular to the field and perpendicular to the velocity. It would be proportional to the charge. An observer on a moving frame could claim that this is a stationary charge. It experiences a force proportional to its charge because there is an electric field at the particle's position. This is the present-day interpretation. Electric and magnetic fields are not absolutes, in that they have a meaning independent of the observer's motion. There are definite rules as to how they transform when being measured by observers on different inertial frames of reference.

You are being given the benefit of information that wasn't available in Maxwell's time. Since the only information that was available then was obtained from experiments on earth, it wasn't known how things would appear as viewed from other frames of reference. The Lorentz force law hadn't been published yet. However, Maxwell knew that magnetic fields exerted forces on wires that were carrying electric currents. In fact, the electric motor had already been invented. Two phenomena that led him to the discovery of electromagnetic waves had to do with electric currents flowing in circuits. These experiments were made possible by the invention of the electric battery in the early 1800s. Two Italian physicists should be given credit for this invention. The first was Luigi Galvani (1737–1798), who placed a piece of copper and a piece of zinc together inside a frog that had been dissected. The twitch of the legs suggested that something was flowing through the nerves of the frog. A few years after Galvani's death, Alessandro Volta (1745–1827) invented the first battery. The electrolyte was sulfuric acid, and the metals were copper and zinc.

Hans Christian Oersted (1777–1851) is usually credited with the discovery that electricity and magnetism are related. In 1820 he was delivering a lecture and closed a switch that established an electric current in a long straight wire. He noticed that the needle of a nearby compass deflected from its previous orientation. This was the first evidence that an electric current could create a magnetic field.

Ampere's Law

There is a similarity in the history of the relationship of electric current to the magnetic field with that of charges to the electric field. A law that is analogous to Coulomb's law is the Biot-Savart law. The analogy is based on the idea that Coulombs Law gives a direct link between the charges and the field while the Biot-Savart Law directly links the magnetic field to the current in the wire. In applying this law to a circuit, it is convenient to treat the wire as having negligible thickness. Contributions to the magnetic field come from each infinitesimal section of wire. An infinitesimal section of wire of length dl and carrying current I is called a current element. It is represented by a vector that points in the direction of current flow. If **r** is the position vector of a point in the field, as measured from this current element, the contribution to the field is given by the formula

$$d\mathbf{B} = \frac{\mu_0}{4\pi} I d\ell \times \mathbf{r}/r^3.$$

In this formula, μ_0 (mu subzero) is a constant that will be discussed later. It depends on the units chosen. This formula is useful for calculating the magnetic field by a steady current in a circuit made up of thin wires. For example, it is a simple matter to calculate the magnetic field at the center of a circular loop of wire of radius R, carrying a current I. All the contributions are in the same direction and the cross product is simple to calculate. The answer is

$$B = \frac{\mu_0}{4\pi} I (2\pi R) \frac{R}{R^3} = \frac{\mu_0 I}{2R}.$$

You should have no trouble in finding the direction of the field.

A more difficult problem is the calculation of the magnetic field at a distance R from a long, straight wire carrying a steady current. The wire is part of a circuit, but the rest of the circuit is assumed to be too far away to matter. The wire extends along the x-axis, as shown in the figure.

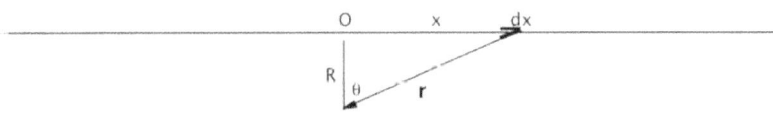

In the figure, the origin, O, is chosen to be directly above the point of interest. According to the Biot-Savart law, **B** is directed into the paper if the current is in the positive x direction. The magnitude of the field is given by an integral:

$$B = \frac{\mu_0}{4\pi} I \int_{-\infty}^{\infty} \frac{R dx}{r^3}.$$

If you have no interest in mathematics, you may skip the rest of this section. However, no new ideas are needed in order to evaluate the integral. It helps to express all the variables in terms of the angle θ. In particular:

$$x = R \tan \theta.$$

Readers who are familiar with calculus can make use of the following equation

$$dx = \frac{R}{\cos^2 \theta} d\theta.$$

(For those readers who would like to derive this result the following hints are given: Express the tangent in terms of the sine and cosine. The derivatives of these two functions can be found in the section on SHM. The calculation is a bit tedious, however.) The remaining variable, r, in the integral can be expressed as

$$r = \frac{R}{\cos \theta}.$$

It is only necessary to note that as x varies from negative infinity to positive infinity, θ varies from $-\pi/2$ radians to $+\pi/2$. The expression for B becomes

$$B = \frac{\mu_0 I}{4\pi R} \int_{-\pi/2}^{\pi/2} \cos \theta d\theta.$$

Since the cosine is the derivative of the sine, this integral can be evaluated immediately. The result is

$$B = \frac{\mu_0 I}{4\pi R} [\sin \frac{\pi}{2} - \sin(-\frac{\pi}{2})] = \frac{\mu_0 I}{2\pi R}.$$

Calculations based on the Biot-Savart law can become extremely difficult, even for simple geometries. It is of limited use.

The French mathematician and physicist Andre-Marie Ampere (1775–1836) is best known for his investigations into the properties of the magnetic field generated by electric currents. He demonstrated that two parallel wires carrying current in the same direction would attract one another. There is a repulsive force if one of the currents is reversed. He established that the force was proportional to the currents, the lengths of the wires, and inversely proportional to the distance between them. Ampere's researches into the properties of the magnetic field led to an important equation known as Ampere's law. It is usually written in terms of an integral that is similar to the one that was used to show that an electrostatic field is conservative. It is illustrated by consideration of the circuit below:

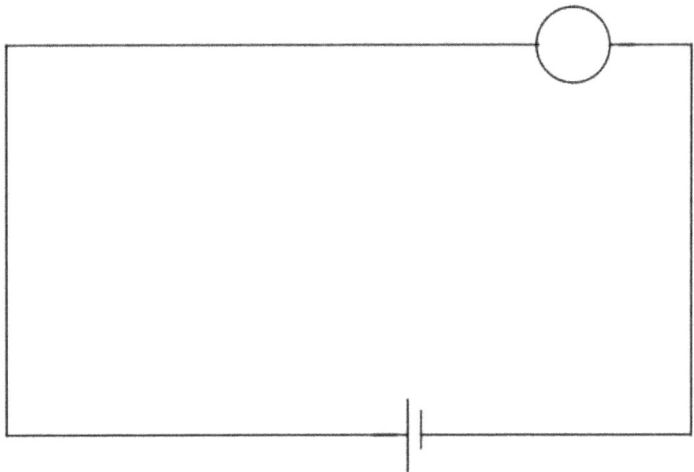

The circle in the diagram represents a device, such as a light bulb. It is needed to regulate the current in the circuit. In terms of the electric current, I, flowing in the circuit, Ampere's law is

$$\oint \mathbf{B} \cdot d\mathbf{r} = \mu_0 I,$$

where the path taken in the integral is an arbitrary closed loop in space. The result of this integration is 0 unless the path surrounds the current in the circuit. In that case, the path and the circuit are like two links in a chain.

Ampere's law can be used to deal with the same problem that was used to illustrate the Biot-Savart law. The problem is to find the strength of the magnetic field at a distance, R, from a long straight wire carrying a current, I. The integral involves a circular path of radius R, with the wire at the center. The magnetic field at each point on this path is in the direction of the path, and its magnitude is constant. The integral is, therefore, B2πR. This leads to the same result as was obtained earlier. Ampere's law is much simpler for this problem. Note that the long straight wire should be considered as a part of a circuit containing a battery. Only the straight section of the circuit is considered close enough to make a contribution to the field.

The Biot-Savart law was considered to be analogous to Coulomb's law in that it specified the field in terms of its sources. In a sense Ampere's law is like Gauss's Law in that it only specifies certain properties of the field. Other factors are needed, such as symmetry, in order to determine the field. Nevertheless Maxwell considered Ampere's law as being more fundamental than the Biot-Savart law. However, Maxwell found it necessary to amend Ampere's law, in order to deal with more general situations than simple circuits.

It is easy to become confused about some of the concepts used in dealing with magnetic fields. Electric currents have been mentioned, but a precise definition of current has not been given yet. It is now known that the electric current in a wire is associated with the motion of electrons. If electrons are moving to the left in a wire, the current is to the right. This is due to the fact that an electron has a negative charge. A current of 1 ampere means that 1 coulomb is passing through each second. This statement implies that the concept of charge is more fundamental than the concept of current. It is thus confusing to the novice that the unit of charge, the coulomb, is defined in terms of the unit of current, the ampere. Although this seems absurd, there is a good reason for doing things this way. If charges are to be determined by measuring the force that one particle exerts on another, the experimental procedure is difficult. Accurate results are almost impossible to attain. Much more accurate results can be obtained by measuring the force on a wire carrying current. The experimental procedure is outlined below.

Since an electric current consists of moving charges, a current-carrying wire will experience a force in a magnetic field. The infinitesimal force on a small section of wire is given by the formula

$$d\mathbf{F} = Id\ell \times \mathbf{B}.$$

Actually, Ampere proposed this law long before Lorentz defined the field in terms of moving charges. The reason for not following an historical approach here is that the subject would be more difficult to understand in that case. It is not hard to show that Ampere's formulation is consistent with that of Lorentz.

Ampere's formula can be used to determine the forces exerted by two long, straight, and parallel current-carrying wires, one on another. For this purpose, it is reasonable to consider one wire as the source of a magnetic field. This magnetic field then produces a force on the second wire. It can then be verified that Newton's third law is satisfied, in that if the role of source were reversed, that the force on the other wire would be of the same magnitude and in the opposite direction. The two wires in the diagram are shown in cross section as circles. The lower one is treated as the source. The currents in this figure are directed into the page. The magnetic field, **B**, arises from the current in the lower wire. The large circle represents the locus of points for which this field has constant magnitude. The direction of the field at any point is tangent to this circle. Making use of Ampere's force law, it can be concluded that the force on a section of length, L, on the upper wire is

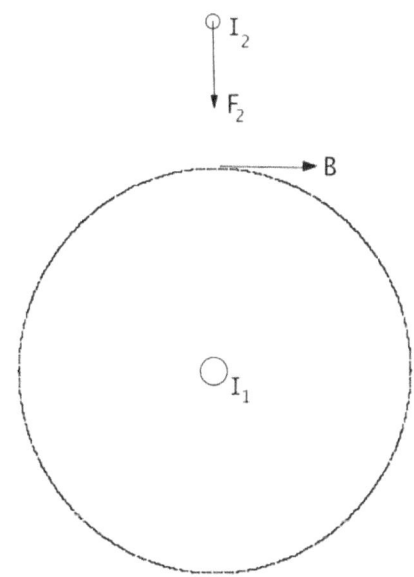

$$F = \frac{\mu_0 I_1 I_2 L}{2\pi r},$$

where r is the distance between the wires, center to center. It is easily verified that the force on the other wire is in the opposite direction from the one shown as **F₂** and is in the opposite direction. In other words, the wires attract one another, and Newton's third law is satisfied.

The experiment just described is used for defining the unit of current in the MKS system of units. This unit, the ampere, is defined by an experiment in which the force on a wire is measured. Two parallel wires are set 1 meter apart, and they are part of a circuit such that the same current passes through each one of them. The ampere is defined as the current that produces a force of 2×10^{-7} Newtons on a meter section of one of the wires. According to the last formula, this leads to the following result:

$$2x10^{-7} = \frac{\mu_0}{2\pi}. \quad \mu_0 = 4\pi \times 10^{-7}.$$

This amounts to a definition of the constant μ_0. It is not necessary to check for units at this point.

The unit of charge in the MKS system, namely, the coulomb, is defined as the amount of charge that passes through a wire carrying a current of 1 ampere in 1 second. The constant that appears in Coulomb's law is then determined by an experiment that measures the force between particles with known charges. The numerical value of this factor is

$$\frac{1}{4\pi\varepsilon_0} = 9x10^9.$$

The reason for writing it in this way is that Maxwell's theory involves the quantity $1/\varepsilon_0\mu_0$, which can be seen to be 9 x 10^{16}. It should be realized that these quantities have units. It will become clearer later what the unit should be. Note that this number is not exact since one of the constants used in its definition is determined by experiment.

Magnetic Monopoles

There is a type of symmetry in the description of electric fields and magnetic fields. Consider the following equations, for example:

$$\oint \mathbf{B} \cdot d\mathbf{r} = 0, \qquad \oint \mathbf{E} \cdot d\mathbf{r} = 0.$$

The first of these equations is Ampere's law in the absence of currents. The second describes electrostatic fields in general. In addition, consider the following equations:

$$\oint \mathbf{B} \cdot d\mathbf{A} = 0, \qquad \oint \mathbf{E} \cdot d\mathbf{A} = 0.$$

The first equation stems from the fact that there are no magnetic monopoles. It is valid in general, while the second is valid in the absence of electric charges. The main difference between the two fields has to do with the fact that electric charges exist, while magnetic monopoles don't.

It is tempting to speculate about the existence of magnetic monopoles. Perhaps an electric field would affect a moving monopole in the same way that a magnetic field affects a moving charge. If moving electrical charges in a wire can create a magnetic field that satisfies Ampere's law, perhaps moving magnetic monopoles could create an electric field that satisfies a similar law. Such a law would be of the form

$$\oint \mathbf{E} \cdot d\mathbf{r} = kI_B,$$

where k is a constant, and I_B represents the current flow of monopoles.

Faraday's Law of Induction

Michael Faraday (1791–1867) was a British experimental physicist. He is to be considered among the greats in the physical sciences. He explored a huge number of electric and magnetic phenomena and authored several books on his researches. Perhaps his most practical achievement was the invention of the electric motor. His most famous contribution to the subject of electromagnetism is his law of induction. Faraday was a strong believer in fields. He pictured electric and magnetic fields as lines emanating from charges and magnets. The fields were strong where the lines were close together. To Faraday, these lines were real.

In order to imagine Faraday's thought processes, consider the following experiment, in which a bar of copper is moved in a magnetic field. Although the bar is neutral, it contains both positively and negatively charged particles. There will be forces on these particles by virtue of the fact that a moving charge, in a magnetic field, feels a force given by $\mathbf{F} = q\mathbf{v} \times \mathbf{B}$. The forces on the negative charges and the positive charges will be in opposite directions. Charges of one type will accumulate on one end of the bar while the opposite type accumulates on the other. An electric field, $\mathbf{E}$, builds up in the bar until $\mathbf{E} = -\mathbf{v} \times \mathbf{B}$. The net force on the charges in the wire is then 0.

It should be noted that the electric field in the conducting bar is not 0. Or is it? It depends on your frame of reference. If you are running alongside the bar as it moves with respect to the laboratory, it is stationary with respect to you. You say that the bar has to be at a constant potential. There is no electric field inside the bar. An observer standing in the laboratory, however, claims there is one. You are both right. It has already been suggested that electric and magnetic fields are not absolutes. They depend on the reference frame.

What if the bar is sliding along metallic rails, as in the figure below?

It is assumed that a uniform magnetic field of magnitude B is directed into the page. Assume also that the rails are good electrical conductors and that the gadget on the right reads the potential difference between the rails (a voltmeter). Charges collect on the rails until the potential difference between the two rails is given by vBL. An electric current, I, flows in the circuit that passes through the voltmeter. There is a force on the rod given by BIL. Somebody, or something, has to push the rod to keep it from slowing down.

No new law of physics is needed to explain the way current can be produced in the circuit. However, such an experiment might have led Faraday to think along the following lines, "A current can be made to flow in a circuit by moving part of it with respect to a magnet. Is it possible to get a current to flow by moving a magnet past a circuit?" There is another way of looking at the situation in the diagram above. The section consisting of bar rails and voltmeter constitutes a circuit. The area of this circuit is changing with time so that the magnetic flux through it is changing with time. The expression for this rate of change is vLB, which is what was obtained for the reading of the voltmeter. Faraday could have wondered if the magnetic flux in a stationary circuit were changing at this rate a similar voltage could be produced.

Faraday experimented with circuits that were fixed in space and examined the effect of a changing magnetic flux. The result was that an electric field was produced that was different from that of an electrostatic field. It is not conservative. In fact, Faraday's law of induction can be written in the following way:

$$\oint \mathbf{E} \cdot \mathbf{dr} = -\frac{d\Phi_B}{dt}, \quad \text{where } \Phi_B = \int \mathbf{B} \cdot \mathbf{dA}.$$

In this equation, Φ is the Greek letter *capital phi* and designates the flux that passes through the area bounded by the path implied by the integral. The minus sign deals with the direction of the field. In the absence of a changing magnetic flux, Faraday's law reduces to the fact that the electric field is conservative. A separate law that says electrostatic fields are conservative is no longer needed. Faraday's law is often applied to circuits, in which case the integral is called the electromotive force, or EMF. Sometimes the notion of EMF is applied to a circuit that contains a battery. However, the meaning of EMF is a bit different in that case. It is difficult to interpret the question of what the integral means if the path in the integral passes through a battery. For a battery, the EMF is just the voltage that exists between its terminals.

Faraday's discovery had an enormous impact. Electricity was no longer a toy for academicians. Generators were now available to produce electricity for industrial use. Together with the electric motor, a whole new world of technology opened up. In order to produce a changing magnetic field, it isn't necessary to make use of moving magnets. The electrical transformer is a device in which a coil of wire is wrapped around an iron ring that is magnetized by a current in a different coil of wire. If an alternating current is passed through such a device, a magnetic field is produced that changes with time. The voltage produced can be of almost any strength desired.

Maxwell's Equations

Electromagnetic theory seemed incomplete to one man. It was the great physicist James Clerk Maxwell (1831–1879) who brought order out of the multiplicity of theories circulating at that time. It was his efforts that helped sort out which of the many laws that existed were fundamental. He was the first person to emphasize that all of electrostatics could be derived from Faraday's law and Gauss's law for electric fields. Maxwell also showed that Gauss's law for magnetism, taken together with Ampere's law, fit all the evidence then known about the magnetic field. There was a problem with Ampere's law in Maxwell's mind, however. It was perfectly adequate for the situation in which steady currents flowed in circuits. What made Maxwell suspect that this law was not quite complete? He must have realized that a steady current in a wire was not the only type of flowing charge. Electric currents didn't have to flow in circuits. He certainly was aware of the similarities between electric and magnetic fields that have already been discussed. Was it possible that these similarities could be pushed even further? Faraday's law states that a changing magnetic field can produce an electric field. Is it possible that a changing electric field could produce a magnetic field? Maxwell's conclusion was that this was actually the case. In order to show this, he first demonstrated that Ampere's law was incomplete.

Consider the circuit shown in the diagram below. The two vertical sections are large metallic plates viewed edge on. They are held apart by some thin insulating material. This arrangement constitutes what is known as a parallel plate capacitor. It is a device for storing electrical charge. When the switch is closed, electrical current flows in the wires connected to the battery indicated at the bottom. Note that no current is flowing through the insulating region between the plates. While the current flows, one plate is accumulating negative charge while the other accumulates an equal positive charge. An electric field builds up in the region between the plates. As long as current flows, there is an accumulation of charge on the plates and a changing electric field between them. Maxwell could see the connection between an interrupted current flow and a changing electric field. Note that the electric field builds up until the potential difference between the two plates is the same as that between the terminals of the battery.

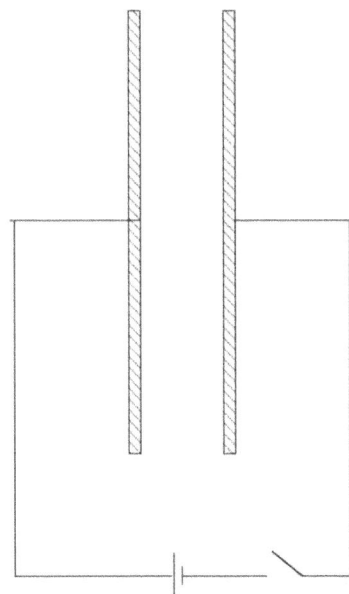

Ampere's law involved an imaginary loop that linked up with a closed circuit. The situation that Maxwell visualized had a gap. There was a break in the current flow...unless...! Maxwell had the brilliant idea that a changing electric flux was equivalent to a current flow. Whenever there was a current in one part of the circuit, there was a changing electric flux in the other. He could calculate the equivalence. The rate at which the charge on a plate was changing was equal to the current in the wire. The electric flux that went from one plate to the other was proportional to the charge, by virtue of Gauss's law. We find

$$\frac{d(flux)}{dt} = \frac{d}{dt}(\int \mathbf{E} \cdot d\mathbf{A}) = \int \frac{d\mathbf{E}}{dt} \cdot d\mathbf{A} = \frac{d}{dt}(q/\varepsilon_0) = I/\varepsilon_0,$$

where I is the current in the wire. Actually, the integral should be performed over a closed Gaussian surface surrounding the charge on one of the plates. In this case, however, almost all the flux is confined to the region between the plates. Maxwell concluded that a changing electric flux was equivalent to a current except for a proportionality factor. He modified Ampere's law in such a way so as to include the possibility of both a current flow and a changing flux. It is now written as

$$\oint \mathbf{B} \cdot d\mathbf{r} = \mu_0 (I + \varepsilon_0 \frac{d\Phi}{dt}),$$

$$\text{where } \Phi = \int \mathbf{E} \cdot d\mathbf{A}.$$

The first integral is over any closed contour in space. The second integral is over an area bounded by that contour. The current is the charge that passes through that area each second. The area is not changing with time. It is fixed with respect to a given reference frame. The only reason the electric flux through it is changing is because the electric field is changing.

If the change in Ampere's law had been Maxwell's only contribution to science, it would have been noteworthy. However, he went much beyond that point. He must have suspected that an electromagnetic field could propagate as a wave. A changing magnetic field could create a changing electric field. This, in turn, could create a changing magnetic field, etc. He suspected he would have to apply these laws of electromagnetism to a microscopic region of space in order to understand if waves were possible. He understood that the wave equation was first derived by examination of an infinitesimal piece of a string. Perhaps it would be helpful to look at an infinitesimal region of empty space.

Consider an infinitesimal box, as shown in the figure below.

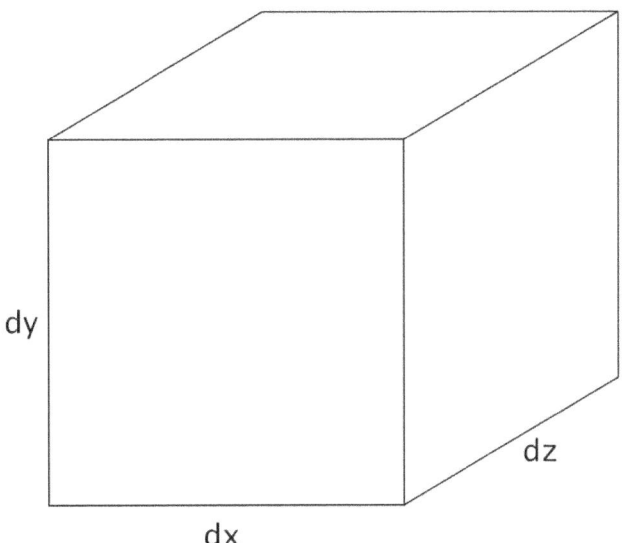

dy

dz

dx

Assume there is an electric field, **E**, present in this region. In order to apply Gauss's law to this box, it is convenient to break **E** into its vector components. In other words, **E** is written as,

$$\mathbf{E} = E_x\mathbf{i} + E_y\mathbf{j} + E_z\mathbf{k}.$$

Each of these component vectors can be treated separately. The first vector is involved in the flux through the left and right faces. The area of each of these is given by dy times dz. If the coordinate of the left face is called x, the coordinate on the right face is called x + dx. If there is no change in E_x in this interval, the inward flux at the left face equals the outward flux at the right face. The net outward flux that arises from the x-component of the field is given by the expression

$$[E_x(x+dx, y, z) - E_x(x, y, z)]dydz.$$

The term in brackets is the change in the field from the left end to the right end. This change can be expressed as the product of the rate of change in the x direction with dx . The outward flux from this contribution is given by

$$\frac{\partial E_x}{\partial x}dxdydz = \frac{\partial E_x}{\partial x}d(volume).$$

It is easy to see that the contributions of the other components have a similar form so that the entire outward flux is given by the expression

$$\left(\frac{\partial E_x}{\partial x} + \frac{\partial E_y}{\partial y} + \frac{\partial E_z}{\partial z}\right)d(volume).$$

Maxwell was able to formulate Gauss's law in terms of the charge density, avoiding the need for integrals. However, since he was interested in a vacuum, the charge was 0. For this case, Gauss's law for electric and magnetic fields takes the following form:

$$\frac{\partial E_x}{\partial x} + \frac{\partial E_y}{\partial y} + \frac{\partial E_z}{\partial z} = 0 = \frac{\partial B_x}{\partial x} + \frac{\partial B_y}{\partial y} + \frac{\partial B_z}{\partial z}.$$

Maxwell applied Faraday's law to an infinitesimal loop, as shown in the diagram below. Use of a right-handed coordinate system requires that the z-coordinate is pointing out of the page. The EMF is positive if it is clockwise. In that case, E_x makes a positive contribution to the EMF on the upper segment and a negative contribution on the lower one.

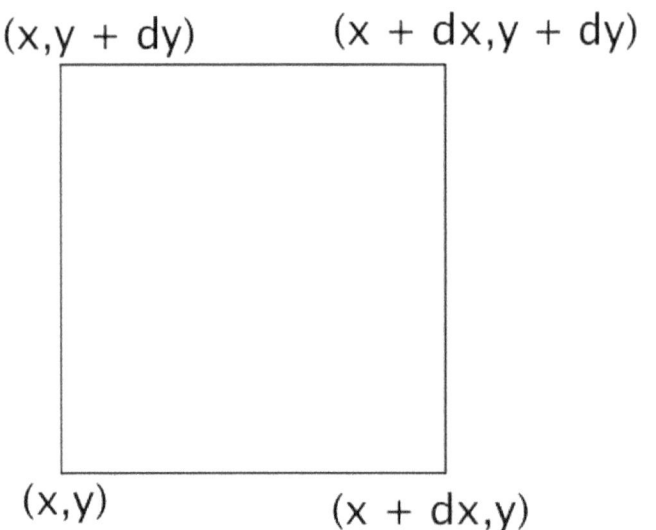

(x,y + dy)   (x + dx,y + dy)

(x,y)   (x + dx,y)

The EMF arising from these two contributions is

$$[E_x(x, y+dy) - E_x(x, y)]dx..$$

This can be written as

$$\frac{\partial E_x}{\partial y} dxdy.$$

The y-component can be treated in a similar way. It should be noted that E_y is associated with a negative contribution to the EMF in the right section. We find the EMF around the loop to be given by

$$(\frac{\partial E_x}{\partial y} - \frac{\partial E_y}{\partial x})dxdy = (\frac{\partial E_x}{\partial y} - \frac{\partial E_y}{\partial x})(area).$$

This is equated to the rate of change of flux through this area. Note that a negative sign is involved with Faraday's law. Because of this sign, a positive EMF is associated with an increasing flux that is directed in the z direction. The rate of change of flux in this direction is given by

$$\frac{\partial B_z}{\partial t}(area).$$

Faraday's law takes the form

$$\frac{\partial E_x}{\partial y} - \frac{\partial E_y}{\partial x} = \frac{\partial B_z}{\partial t}.$$

There are two more equations associated with Faraday's law. These are

$$\frac{\partial E_y}{\partial z} - \frac{\partial E_z}{\partial y} = \frac{\partial B_x}{\partial t}, \quad \frac{\partial E_z}{\partial x} - \frac{\partial E_x}{\partial z} = \frac{\partial B_y}{\partial t}.$$

It is only necessary to replace x by y, y by z, and z by x in the earlier equation to obtain these results.

Ampere's law, as modified by Maxwell, is similar to Faraday's law in a region devoid of currents. The roles of E and B are interchanged for the two laws. The main difference is that the RHS of Ampere's law is positive, and ε_0 and μ_0 must be included. The following equations result:

$$\frac{\partial B_y}{\partial x} - \frac{\partial B_x}{\partial y} = \varepsilon_0 \mu_0 \frac{\partial E_z}{\partial t}, \quad \frac{\partial B_z}{\partial y} - \frac{\partial B_y}{\partial z} = \varepsilon_0 \mu_0 \frac{\partial E_x}{\partial t}, \quad \frac{\partial B_x}{\partial z} - \frac{\partial B_z}{\partial x} = \varepsilon_0 \mu_0 \frac{\partial E_y}{\partial t}.$$

These differential equations are known as Maxwell's equations for empty space.

Electromagnetic Waves

Maxwell sought solutions to the equations he had just developed. Try to imagine his thought process. The differential equations that have been obtained are very complicated in that they involve six quantities that are functions of three spatial coordinates and the time. When faced with such a problem most scientists make simplifying assumptions so as to deal with special cases. Can a solution be found, for example, in which the electric field components E_y and E_z are both 0? In that case, Gauss's law for the electric field would be

$$\frac{\partial E_x}{\partial x} = 0.$$

This implies there is no dependence of the electric field on x. Application of Faraday's law to this situation yields

$$\frac{\partial E_x}{\partial y} = \frac{\partial B_z}{\partial t}, \quad \frac{\partial B_x}{\partial t} = 0, \quad \frac{\partial E_x}{\partial z} = -\frac{\partial B_y}{\partial t}.$$

The second equation indicates that the component of the magnetic field parallel to the electric field doesn't vary with time. Such a static field isn't of much interest. It might be due to some permanent magnet in the vicinity, but it wouldn't have any bearing on any disturbance. Thus, the magnetic field in any wave would have to be perpendicular to the electric field. It is consistent then to assume that the magnetic field is in the y direction, thus, $B_x = B_z = 0$. The equations simplify further. Faraday's law now says

$$\frac{\partial E_x}{\partial y} = 0.$$

The rate of change of the electric field with y is 0. The electric field depends only on the z-coordinate and on the time.

Gauss's law for the magnetic field now is reduced to

$$\frac{\partial B_y}{\partial y} = 0.$$

Moreover, use of Ampere's law yields

$$\frac{\partial B_Y}{\partial x} = \varepsilon_0 \mu_0 \frac{\partial E_Z}{\partial t} = 0.$$

The reason is that E_Z is 0 and doesn't depend on the time. Thus, the magnetic field also depends only on z and t. The only equations that need to be considered are

$$-\frac{\partial E_X}{\partial z} = \frac{\partial B_Y}{\partial t}, \quad \text{and} \quad -\frac{\partial B_Y}{\partial z} = \varepsilon_0 \mu_0 \frac{\partial E_X}{\partial t}.$$

These equations demonstrate again the similarities between the electric field and the magnetic field. The rate of change of one field with respect to time depends on the rate of change of the other with respect to position.

It would be nice to have one equation for each field that depended on space and time. Instead we have two equations that make one field dependent on another. In order to get a single equation for the electric field, it is necessary to differentiate these expressions. The first is differentiated with respect to z, while the second is differentiated with respect to t. Each equation will then involve a second derivative of the magnetic field with respect to both z and t. It is asserted, without proof, that

$$\frac{\partial^2 B_Y}{\partial z \partial t} = \frac{\partial^2 B_Y}{\partial t \partial z}.$$

In other words, the order in which the differentiation is performed doesn't matter. This assumption leads to the final result, namely,

$$\frac{\partial^2 E}{\partial z^2} - \varepsilon_0 \mu_0 \frac{\partial^2 E}{\partial t^2} = 0.$$

Note that the subscript, x, on E is no longer needed, as there is no other component of the field being considered. This equation is exactly like the wave equation that was found for waves on a string, namely,

$$\frac{\partial^2 y}{\partial x^2} - \frac{1}{c^2}\frac{\partial^2 y}{\partial t^2} = 0.$$

Recall that c is the velocity of wave propagation. Maxwell had shown that the electric field he had imagined satisfied the wave equation! The velocity of the electromagnetic wave was given by the expression

$$c = \frac{1}{\sqrt{\varepsilon_0 \mu_0}}.$$

It was already shown that the numerical value of the square of this quantity was 9 x 10^{16} . In the MKS system of units, speeds are measured in meters per second. Maxwell found that the velocity of the wave he was proposing had the value of

$$c = 3.0 \times 10^8 \, \text{meters/second.}$$

Maxwell published his theory in 1861. It turned out that the speed of electromagnetic waves was the same as the speed of light. Accurate measurements had been made just ten years earlier. The fact that light was a wave was not a surprise by then. It had been demonstrated by Thomas Young (1773–1829) at the beginning of the century. Nevertheless, Maxwell's discovery that light was an electromagnetic wave had a tremendous impact on the scientific community, as well as the world at large. One of the most mysterious phenomena, the propagation of a light wave through empty space, was now related to electricity and magnetism. A new world had opened up. Electricity and magnetism had emerged from humble beginnings. Because of the work of people like Faraday and Ampere, it had entered the realm of technology. Now it became clear that everything we could see came to us through electromagnetic waves. It was just a matter of time before the technology developed that would produce all types of electromagnetic radiation from radios broadcasts and microwaves to X-rays. Today, electromagnetic waves are used to communicate, to operate remotes, to navigate. The talent of Maxwell is amazing.

The simplifying assumptions that showed that electromagnetic waves are possible are not too restrictive. The fact that the principle of superposition applies to solutions of the wave equation allows for the construction of a huge variety of waves of the type just considered. The important feature of these waves has to do with the fact that both the electric and the magnetic fields are perpendicular to the direction of propagation. Such waves are said to be transverse. In the derivation of the wave equation, it was assumed that only one direction of the electric field was involved. Such a wave is said to be linearly polarized. In the case of light emanating from a flame or an incandescent source, the direction of the electric field varies in a somewhat chaotic fashion. Polarized light can be produced in a variety of ways. You can use a polarizing filter that absorbs one of the components of the light wave and transmits the perpendicular component, for example.

It has been suggested that the various laws on which Maxwell based his theory were valid in all inertial reference frames. This was not the view held by scientists in 1861. They were laws that were based on experiments performed on earth. How could they be valid on a planet that was moving with a substantial velocity with respect to the earth? If electromagnetic waves are traveling with one speed with respect to us, wouldn't they be travelling at another speed with respect to a moving frame of reference?

Old questions were raised by the new electromagnetic theory. If there is a wave, there must be a medium. That was the thinking of almost every scientist, including Maxwell. The medium came to be named the ether. Scientists were concerned whether the ether was fixed in relation to the sun or possibly carried along by the earth. Maxwell based his theory on laws that seemed to be valid on earth. Is the earth special? Which of these laws would not be valid on Mars? If these questions were asked, they weren't answered in the nineteenth century.

It seems that whenever a scientific discovery is made, it raises as many questions as it answers. Is it possible to determine how fast the planet earth is moving relative to the ether? The earth is traveling in one direction in January and almost in the opposite direction in July. Shouldn't there be a difference in the velocity of the ether with respect to us as determined by measurements taken six months apart? Two famous experimental physicists, Albert A. Michelson (1852–1931) and Edward Morley (1838–1923), tried to find this difference. You can access their experiments on the Internet by a search entitled "Michelson Morley." Although the precision of their apparatus was deemed sufficient to detect a smaller difference than expected, nothing was found. Scientists speculated on whether part of their apparatus might have shrunk when moving in one direction or another with respect to the ether. Nobody could give a satisfactory answer as to why such a thing might happen. It was puzzling. Even if the ether were being dragged along with the earth, experiments would have shown that. It will be seen that Albert Einstein (1879– 1955) was the first person to give a satisfactory answer to these questions. Although he sometimes claimed to be unaware of the Michelson Morley experiment, he was vague on this point. Einstein biographers point out that he had been informed of these experiments, but he didn't rely on them. From the time he was a teenager, he couldn't imagine the possibility of catching up to a light beam and trying to visualize the corresponding static electromagnetic field.

Part 3
Einstein

Chapter 11
Nineteenth-Century Physics

In order to appreciate the contributions made to science by Albert Einstein, it is helpful to consider first what was known when he came upon the scene. Physical theory before the introduction of relativity and quantum theory is known as classical physics. The more recent work is called modern physics, even though it is now over a hundred years old. In the centuries following Newton's laws of motion and of gravitation, much effort was expended in applying these laws to all sorts of phenomena. Hydrodynamics and acoustics, for example, involved the application of Newton's laws to liquids and gases.

Two mathematicians, Joseph Louis Lagrange (1736–1813) and William Rowan Hamilton (1805–1865) reformulated Newton's laws into a form that made them more suitable for solving complex problems. Instead of relying directly on forces, these formulations treated the potential energy and the kinetic energy of systems as fundamental quantities. Force, being a vector, is much more difficult to handle than energy. Both relativity and quantum theory are related to these classical adaptations of Newton's laws. Modern physics relies heavily on the work of Lagrange and Hamilton.

The wave nature of light had already been discovered before Maxwell published his equations. Thomas Young (1773–1829) demonstrated this aspect of light early in the nineteenth century. You are encouraged to read about his famous double-slit experiment on the Internet. Shortly after the discovery that light was a wave, Joseph von Fraunhofer (1787–1826) invented the spectroscope, which he used to study the spectrum of the light emitted by the sun. It made use of a diffraction grating, consisting of closely spaced slits, which enabled him to determine the wavelength of the spectral lines. Fraunhofer discovered that certain discrete wavelengths of light were missing from the spectrum. Spectroscopy was later applied to the study of gases. The missing lines turned out to be characteristic of certain elements in the sun's atmosphere. An interesting sidelight has to do with the fact that the element helium was discovered on the sun before it was discovered on the earth! These discoveries by Fraunhofer turned out to be important clues to the understanding of atoms in the twentieth century. Atomic theory, in turn, played an important role in the development of quantum theory.

In the nineteenth century, many physicists didn't believe in the existence of atoms. The English chemist John Dalton (1766–1844) was among the first to make serious arguments in favor of atomic theory. Although the theory explained how chemical elements combine, it gave no idea as to the size of atoms or how to detect them directly. There were some theorists who attempted to explain the properties of gases by means of kinetic theory. This theory applied Newtonian physics to the motion of atoms and molecules in order to explain how the pressure of a fixed amount of gas was related to its volume and temperature. Maxwell was one of these theorists. The Austrian physicist Ludwig Boltzmann (1844–1906) also became famous for his use of probability theory in this study. Both Maxwell and Boltzmann share credit for their studies of the distribution of velocities of the molecules in a gas.

One of the major achievements of the nineteenth century came from the study of thermodynamics. This was a brand-new field, stimulated in part by the development of the steam engine and other power sources. What is now called the conservation of energy is more formally known as the first law of thermodynamics. The English physicist James Prescott Joule (1818–1889) showed that a loss of energy by a mechanical system left a record of some sort. For example, suppose a water bath is used to cool a mechanical system. If the system loses a certain amount of mechanical energy, the water bath will end up at a definite temperature that only depends on its initial conditions. In other words, the changes in the state of the water bath provide a record of the energy lost by the mechanical system. This can be regarded as an exchange of energy between the mechanical system and the bath. No longer is it necessary to restrict energy to mechanical systems. It can reside in batteries, electromagnetic waves, food, and thermal baths, for example.

The second law of thermodynamics is just as famous as the first. It is attributed to the French physicist Sadi Carnot (1796 - 1832). The law is subtle and has been misstated in print many times. It deals with the degradation of the energy that occurs when some process takes place in which there is an exchange of energy. In the example given, it is assumed that energy has flowed from a mechanical system to a water bath. The energy cannot be put back into the mechanical system without some agent leaving the rest of the "world" changed somehow. There is no catalyst or agent that can achieve this without altering the environment.

The second law often takes the form of a statement to the effect that it is impossible to do such and such. It is amazing that such a statement could lead to the construction of a quantity called entropy. This is a quantity that can be assigned to a system in such a way that the total entropy change of a system, plus its surroundings, goes up in any spontaneous process. As another example, consider a large chamber that is divided into two compartments. One compartment contains air at high pressure. The other compartment is empty. An opening is now made in the wall dividing the two compartments. After a while, the air is calm, and things settle down. The entropy of the new system is higher than the old. The gas will never divide up into two again. Yes, it is mathematically possible, but you can wait many eternities before it will even approximately divide into two regions of measurably different pressure again. To all intents and purposes, it is impossible.

The second law of thermodynamics has attracted the attention of many philosophers as well as many physicists. If entropy is always increasing, then why have we inherited this universe in such a low state of entropy in the first place? Is it possible that somewhere in the universe some unknown law is at work, reducing its entropy?

Statements are often made that the second law of thermodynamics implies that all systems must grow more chaotic with time. Some people even argue that evolution from simple systems to more complex systems is impossible because of this law. These are incorrect conclusions. The entropy of a system can decrease as long as some environmental change is also occurring at the same time. It is the total entropy of the system plus the environment that is increasing. When a lake freezes over, its entropy decreases. The process involves an exchange of energy with the atmosphere. The entropy of the atmosphere increases by an amount greater than the loss by the lake.

The subject of heat and thermodynamics led to a new subject of investigation called statistical mechanics. This is a theoretical area in which the laws of probability are used in modeling a system. It differs from kinetic theory in that it is much less detailed in the models it uses. In statistical mechanics, a physical system is often characterized as being composed of a huge number of identical subsystems. These subsystems are randomly exchanging energy and momentum with one another. Many important properties of physical systems can be deduced by means of this type of analysis. Even after the introduction of quantum theory in the twentieth century, the subject of statistical mechanics is still important.

Maxwell's Demon

Maxwell mused long and hard about the second law of thermodynamics. The statement that something was impossible challenged him. He imagined a kind of cylinder with adiabatic (perfectly insulating) walls. The cylinder had two compartments that were also separated by an adiabatic wall. In this wall there was a hole that was big enough to let a molecule of gas slip through. Maxwell imagined that there was the possibility that a small enough "demon" could conceivably act as a gatekeeper at this hole and allow only certain molecules to pass through. There was a gas in both compartments, and molecules were constantly approaching the hole from both sides. The demon could presumably make the left compartment hotter than the right one by only allowing fast molecules from the right through the hole, while allowing only slow molecules approaching from the left to pass through.

If Maxwell's demon could have accomplished the separation of a hot gas and a cold gas from one that was at an intermediate temperature, it would have succeeded in lowering the entropy of the universe. Is this impossible? This question will not be answered here. It has fascinated physicists for a long time. Richard Feynman (1918–1988), a charismatic, ingenious, Nobel Prize–winning theoretical physicist, spent a whole semester on this single question, while teaching a course in statistical mechanics. One problem that the demon faces relates to how the information about the speed of the molecules is to be gained. Does the entropy of the universe go up in order for any of us to gain information? There definitely seems to be a connection between information and entropy. The recording of information seems to be an irreversible process. Molecular chaos is a form of misinformation. If a detectable pattern could emerge in molecular motion, it could be exploited to make use of the energy that becomes available. To this day, Maxwell's demon has stimulated much thought, and controversy.

An important landmark of the twentieth century was the discovery of the electron. In a sense it was Thomas Edison's invention of the incandescent light bulb that led to this discovery. Edison's first successful light bulb made use of a carbon filament. It had a problem in that carbon was evaporating and getting deposited on the walls. In order to reduce the darkening of the walls, Edison experimented with a light bulb into which an extra electrode was placed. In essence he constructed a vacuum tube. It was found that a current could pass through the vacuum in the tube, but in only one direction. It is called the Edison effect. It was deduced that particles with a negative charge were being emitted from the hot filament. Later experiments by men such as J. J. Thomson (1856–1940) modified the tube into what is called a cathode ray tube. It is similar to the picture tube used in old TVs. In such a device, the negative particles are accelerated toward a positive terminal but allowed to pass through a hole in the terminal forming a beam of charged particles. By means of electric and magnetic fields, the beam could be deflected in controlled ways. Thomson was able to deduce the ratio of the charge of this particle to its mass. It was the discovery of the electron. This work, which was published in 1897, led to the award of the Nobel Prize in 1907.

The cathode ray tube led to another important discovery. Wilhelm Roentgen (1845–1923) received the first Nobel Prize in physics for his discovery of X-rays. He had been working with cathode rays. His experiments involved accelerating potentials of 1,000 volts or more. When energetic electrons impinged on an aluminum target, some invisible radiation caused a phosphorescent material to glow. It was later discovered that Roentgen rays, or X-rays, are electromagnetic waves of extremely high frequency. They are widely used in the medical profession because of their penetrating power. They have also been very useful to the scientific community. Because their wavelengths are comparable to the atomic spacing in crystals, X-ray diffraction by crystalline materials proved to be an excellent method for the determination of the atomic arrangements in solids. It should be mentioned that most physicists didn't believe in atoms or molecules in the nineteenth century. This was true in spite of the atomic and molecular basis of the kinetic theory of gases. X-ray crystallography wasn't developed until many years after the discovery of X-rays in 1895.

Henri Becquerel (1852–1908), a French physicist, was influenced by Roentgen's discovery. He knew that X-rays caused certain materials to fluoresce. The materials would then emit radiation for some time after the X-ray source was taken away. Becquerel had a material, a compound of uranium, which he had known to fluoresce. He was going to use sunlight to activate the material, but the day turned out to be overcast. He wrapped the material in paper and placed it in a drawer on top of some photographic film. He later discovered that the film had been exposed. In this way he was led to the discovery of radioactivity. This discovery came one year after the discovery of X-rays.

Marie Curie (1867–1934) and her husband Pierre carried out further studies in the physics and chemistry of radioactive materials. Marie, who was born in Poland and later became a French citizen, was the first person to receive two Nobel Prizes. One, in physics, which was shared with her husband and with Becquerel, was awarded in 1903. The other was in chemistry. It was awarded in 1911.

There were three different types of radiation associated with radioactive materials. They were given the names associated with the first three letters of the Greek alphabet: alpha, beta, and gamma. The first of these, alpha radiation, consists of positively charged particles. These particles are now known to consist of the nuclei of helium. They are about four times as massive as hydrogen nuclei and carry twice the charge. Beta radiation is nothing other than cathode rays. In other words, they are electronic emissions. They differ from typical cathode rays only in that their energies are typically much higher. You would have to put 1 million volts across a cathode ray tube to get such energetic electrons. The rays in the last category, gamma rays, are electromagnetic. They have wavelengths of various sizes, but typically they are shorter than X-rays. The scattering of alpha particles by a thin metallic sheet led to a revolution in physics and chemistry.

Chapter 12
The Miracle Year

Two mind-boggling ideas were thrust upon the physics community in the first decade of the twentieth century. The first of these ideas is now called quantum theory. It had modest beginnings. Max Planck (1858–1947) was attempting to explain the distribution of different frequencies in the radiation emitted by a hot body. In particular, the body with which he was dealing was called black and the radiation was called black body radiation. To physicists, the word *black* in this theory has to do with the fact that the body is presumed to absorb all the radiation that falls on it. It is an ideal in a sense, in that thermodynamics predicts that if such a body were to exist, it would be the strongest emitter possible. If such a body were a poor emitter of radiation, it would be possible to break the second law of thermodynamics. The argument goes as follows: Suppose it is possible for two warm bodies to exchange energy by means of thermal radiation only. You might visualize them as being in some enclosure with perfectly reflecting walls so that no energy could leak away. If black bodies were poor emitters, it is conceivable that even when two bodies were at the same temperature, one body would absorb more energy than it emitted. There would arise a situation in which two bodies were initially at the same temperature, but as time went on, one would get hotter than the other. This is not possible according to the second law of thermodynamics.

A small window to a hollow enclosure is a good approximation to a black body. Any light that enters the window would bounce around many times, getting absorbed over and over in such a way that almost no light would emerge through the same window. In that sense, the window would be black. You may have looked through the window of a kiln or a furnace when all the objects inside are glowing red from the heat. If all the objects have reached the same temperature, you will not be able to pick out any objects with your eyes. They have all merged, in the sense that you are only seeing the radiation associated with black bodies. Those objects that have surfaces that are white when cool are reflecting the radiation that they don't absorb. The study of black body radiation is essentially the study of the electromagnetic waves inside an enclosure at a uniform temperature.

The radiation emanating from hot bodies is familiar, in the sense that white-hot is known to correspond to a higher temperature than red-hot. Classical theories could not explain the actual frequency distribution of this radiation. Planck made an assumption that sounded strange, even to him. It was equivalent to imagining that a black body consisted of a collection of simple harmonic oscillators that had a very peculiar property. He assumed that each oscillator could only absorb energy in chunks. In classical physics, it is assumed that the amplitude of vibration of a simple harmonic oscillator could be anything. In other words, all energies are possible for an ideal spring and mass. Planck saw that this model didn't fit the observations. He could only make the oscillator model fit the observed data by making the assumption that only certain energies are possible. If an oscillator had a natural frequency, v, (the Greek letter *nu*), the only possible energies that such an oscillator could have seemed to be given by the formula

$$E_n = nhv,$$

where n is an integer $(0,1,2,\cdots)$, and h is some constant. The constant h is now called Planck's constant, and has the numerical value of 6.626×10^{-34} in the MKS system of units.

When Planck hit upon this model, he was able to fit the frequency distribution of Black Body Radiation for all temperatures for which it had been measured. The theory might have been strange, but it was very successful.

The year of Planck's theory was 1900. The physics community, as well as Planck himself, regarded the constant h as a mathematical term of no great significance. Perhaps it had something to do with the oscillators that were assumed to be interacting with the light. Just a few years later, Einstein made the claim that light also behaved as though it came in chunks. In other words, it acted as though it were made up of particles. It was 1905, often called Einstein's Miracle Year, the year in which he published four remarkable papers. Before that year, he had been an obscure researcher, having published a few papers that had been barely noticed. He had not earned his PhD degree yet. He couldn't even get a teaching assistantship. His university professors had little use for him because of his rudeness and unwillingness to defer to authority. Because of that, he could not get any references from them for a teaching position. He was lucky to land a job at the Swiss Patent Office, thanks to the efforts of a close friend.

Each of Einstein's four papers in that miracle year could have been considered worthy of a Nobel Prize. In fact, he did receive the prize sixteen years later for one of them. However, it was not for his most famous theory, the special theory of relativity. This was the subject of his last paper that year. The prize was for his paper on the photoelectric effect.

The photoelectric effect has to do with the emission of electrons from certain metals when the surface is illuminated. There was something strange about the phenomenon in that the energy of the electrons that were emitted had nothing to do with the brightness of the light. The expected behavior was that the light would carry energy and heat up the region. Gradually electrons would get enough energy to break through any barrier that confined them to the interior of the metal. High-intensity light should heat the material more rapidly and provide the electrons with more energy. The experiments of the German physicist Phillip Lenard (1862–1947) in 1902 showed, however, that the energy of the ejected electrons depended only on the frequency of the incident light waves. There was no warm-up period. Electrons were emitted right after the metal was illuminated. The intensity of the light only affected the rate at which electrons were emitted. Einstein was able to explain this behavior by taking Planck's paper seriously. Although he was fully aware of the wave theory of light, Einstein claimed that light also had corpuscular properties. He referred to these bundles of energy as light quanta. We now call them photons. In Einstein's model, each photon carries a specific energy, dependent only on the frequency of the light. This energy is the product of the frequency with Planck's constant. Einstein assumed that when a photon is absorbed, it transfers all of its energy to a single electron. Electrons are free to move around inside metal, but it is assumed that there is a potential barrier that keeps them confined to the interior. There is a minimum energy that must be provided by the light quanta for even the most energetic electrons to escape. This energy is called the work function of the metal. Thus, Einstein's theoretical model predicted no electron would escape unless the photon had an energy equal to the work function. Therefore, there was a minimum frequency necessary for light to cause photoemission. If the frequency were higher than this minimum, it would be possible for electrons to escape. The number escaping per second would be proportional to the intensity of the light. However, the maximum energy possible for escaping electrons would be dependent only on the frequency. The theory led to the formula

$$E_{max} = h\nu - \phi,$$

where phi is the work function of the metal. There were many who didn't take this paper seriously until ten years later when the noted American experimental physicist Robert Andrews Millikan (1868–1953) confirmed this equation by means of a beautiful experiment. Millikan received the Nobel Prize in 1921 for his work on the photoelectric effect as well as for the measurement of the electronic charge a few years earlier. This latter experiment also specified the electron's mass, because J. J. Thomson had already measured the charge to mass ratio years earlier.

You might try to imagine the state in which Einstein's photoelectric theory left the field of physics. How can the wave nature of light be reconciled with its corpuscular behavior? Could Huygens and Newton both be correct in their description of light? It would take years before anyone could make these two different views of light compatible.

The evolution of quantum theory will be sketched out later. It is extremely complicated in that it is still a theory that doesn't fit anyone's intuition, and there have been many interpretations as to what its implications might be. However, nearly every physicist is in agreement that it is a very successful theory, because all of its experimental predictions have been confirmed.

Einstein's special theory of relativity is taken up in the next section. His other two papers in his miracle year finally convinced most physicists of the existence of atoms. Up until then, there was no evidence of the actual size of any atom. Einstein showed how the viscosity of liquids could be used to calculate atomic sizes. He also showed that the motion of pollen suspended in water could be explained by molecular bombardment. This latter motion, called Brownian motion, had been known for a long time, but it had no simple explanation before Einstein put the matter to rest.

The Special Theory of Relativity

The special theory of relativity deals specifically with inertial frames of reference. That is why it is called special. Recall that there are an infinite number of such frames, each moving with constant velocity with respect to one another. It is necessary to look out a window to see whether the frame is in motion with respect to something else. This was Einstein's basic assumption. It was clear that Newtonian physics was valid in all such frames, but it had been thought that Maxwell's equations implied something else. If an electromagnetic wave were used as part of an experiment, wouldn't that be a way of telling whether that laboratory was moving with respect to the earth. After all, the wave couldn't be moving at the same speed with respect to the lab as it was with respect to the earth under these circumstances. Could it? Einstein made the claim that all inertial frames were equivalent. There was no way of detecting the motion of one frame with respect to any other by any sort of experiment performed within that frame. In other words, not only were Newton's laws governing the motion of particles the same in all such frames, so were the laws governing anything else. He claimed that the speed of light is the same in all inertial frames.

How can that be? No matter how fast the wind is moving with respect to an observer on the earth, it is standing still with respect to a person floating in a balloon that is being carried by the wind. The speed of anything always seems to depend on the motion of the observer. Only a genius or an idiot could claim otherwise. Einstein certainly wasn't stupid. He had been thinking hard about how to reconcile his two premises before the Aha! moment came to him. The concept of speed involves the distance covered in a certain length of time. Could it be that the distance between two points in space, or the time interval between two events, depended on the state of motion of the observer? Perhaps both of these quantities did.

Newton had assumed that time was absolute. In his description of motion, it was sufficient to denote the position of a particle at a specific time. There was no question as to which clocks were being used to measure the time. Since Newton's laws worked so well, no scientist ever questioned his assumption. Time was a given. Einstein asked himself, "How do we know whether two events took place at the same time?" If the events took place far from an observer, it was usually necessary to make use of a light signal to determine the exact time of the event. For example, suppose an astronomer notes a volcano erupting on a distant planet. She notes the time on a clock in the observatory. To specify the exact time at which the eruption took place, it is necessary to allow for the time it takes a light signal to reach the observatory from the location of the eruption. After such corrections are made, it is an assumption that two events that took place at the same time were simultaneous. In other words, it was thought that everyone would agree that neither event occurred before the other. Astronomers on the earth are in general agreement as to whether two events took place simultaneously. Einstein wondered whether an observer on a planet in rapid motion with respect to the earth would also agree that the two events were simultaneous. He conjured up a thought experiment. He imagined himself standing on the ground next to a passenger train as it came whizzing by. He then imagined that two lightning bolts struck the train simultaneously just as the midpoint of the train was coming by. One bolt hit the front of the engine, and the other hit the rear of the caboose. The events were simultaneous as far as Einstein was concerned because the signals arrived at the same time, and it was easy for him to deduce that he was standing next to the midpoint of the train when the events took place.

How would the events appear to a passenger seated at the train's center? While the light signals were approaching, she would have moved forward. The signal coming from the engine would reach her first. Since she considers herself to be on an inertial frame, equidistant from the engine to the caboose, she would assume that lightning hit the engine first. By Einstein's assumptions that all inertial frames were equivalent, it is clear that not everyone would agree that two events were simultaneous. It follows from such arguments that clocks that are moving with respect to one another cannot be synchronized. The time at which an event takes place depends on which frame of reference is being used. It should be emphasized that it makes no sense to claim that one set of clocks is better than another. All inertial frames are equivalent in Einstein's view.

The Lorentz Transformation

The discussion so far has been qualitative. Einstein's theory is concerned with certain rules of transformation, giving the relation between coordinate systems, associated with two different reference frames. Consider the train of the preceding section. Let the Cartesian coordinate system that is embedded in the train be called the primed coordinate system. The position coordinates within this frame are x', y', z' (read this as x prime, etc.). The unprimed coordinates are embedded within the stationary frame on the ground. (Of course, it is only stationary with respect to the earth.) It is convenient to choose both x and x' to be parallel to the horizontal track; y and y' can then be chosen to be horizontal and across the tracks, while z and z' correspond to the vertical direction. Time is no long-er the same for all observers so that it is necessary to include a t and t'. There are, therefore, one time and three spatial coordinates to relate to one another.

The origin of the spatial coordinate systems can be chosen arbi-trarily, as can the zeros of time. It is convenient to assume that the spatial origins coincide at $t = t' = 0$. In that case, it turns out that y = y' and z = z' at all times. This is the same relation, the so-called Gali-lean transformation, that existed before the theory of relativity. In other words, if the track is headed in an easterly direction, both an observer on the train and an observer on the ground would agree as to how far north of the track an event took place. The same is true for the altitude of an event.

If this were not the case, a paradox would be encountered. Sup-pose that the size of a moving object were to grow in a direction that is perpendicular to its velocity vector. A passenger on the train would then see the tracks moving and therefore getting further apart. If the train were to go fast enough, the conclusion would be that the train would fall between the tracks. An observer on the ground would conclude that the axle holding the wheels would grow longer and the rails would end up between the wheels. Both observers couldn't be right. A contradiction would arise if a con-traction had been assumed instead of an expansion. The conclusion is that y = y' and z = z'. In this respect, the relativistic theory agrees with the classical one.

Consider the Galilean transformation (pre-relativistic) involv-ing x and x'. This is given by either of the following equations:

$$x = x' + vt, \qquad x = x' + vt'.$$

This is an expression that indicates that the x coordinate of a seated passenger on the train is increasing with time. In these equations, use has been made of the Newtonian assumption that

$$t = t'.$$

The relativistic equations are

$$x = \frac{x' + vt'}{\sqrt{1 - v^2/c^2}}$$

and

$$t = \frac{t' + vx'/c^2}{\sqrt{1 - v^2/c^2}}.$$

In order to show that there is nothing favored about either one of these systems, it is necessary to show that the equations for the primed quantities are the same as those for the unprimed ones, except for the sign of the velocity. After all, if the train is moving eastward with respect to the station, then the station is moving westward with respect to the train. In order to display that this is the case, it is helpful to write the equations in the following form:

$$x = \gamma(x' + vt'), \quad t = \gamma(t' + vx'/c^2), \quad \gamma = \frac{1}{\sqrt{1 - v^2/c^2}}.$$

You should note that these equations imply that v must be less than c, the speed of light. Also the factor, gamma, is a number greater than 1. The variable t' can be eliminated from these equations by multiplication of the second equation by v and subtracting this new equation from the first. This leads to

$$x - vt = \gamma(x' - v^2 x'/c^2) = \gamma x'(1 - v^2/c^2)$$

$$= \gamma x' \frac{1}{\gamma^2} = x'/\gamma,$$

so that

$$x' = \gamma(x - vt)..$$

In a similar fashion, the coordinate x' can be eliminated. It is found that

$$t' = \gamma(t - vx/c^2).$$

These same equations had been published by Hendrik Lorentz (1853–1928) just a few years before Einstein's famous article on relativity. They are known as the Lorentz transformations. Lorentz was a famous Dutch physicist who eventually won the Nobel Prize for other work he had done. His famous force law for moving charges has already been mentioned. The essential difference between Einstein's theory and that of Lorentz has to do with the fact that Einstein treated all these inertial frames as being equivalent. He didn't believe there was an ether that singled out one reference frame as being preferable. Lorentz was convinced that an ether existed so that there was one true frame. He believed that the meter sticks and clocks in other frames had been altered by their motion with respect to the ether. Even after Einstein's theory turned out to be the better one, it wasn't possible to convince Lorentz that there was no ether. Prejudices die hard.

Lorentz actually believed that motion through the ether shrank meter sticks if they were pointed in the direction of motion. Is there an actual compression? It doesn't make sense to think of such a phenomenon occurring if there is no such thing as an ether. All frames are supposed to be equivalent. In which frame would the meter stick be normal? How is the length of a moving meter stick measured anyway? You can't use another moving meter stick if they are both affected in the same way. Einstein provided two answers to the question. One answer involves the notion of simultaneity. If the position of the two ends of the meter stick are recorded simultaneously, a record is left of two fixed points in the "stationary" frame of reference. The distance between these two points can be defined as the length of the moving meter stick. Alternatively, the knowledge of the speed of the meter stick can be used to determine its length. All you need to do is to record the time it takes the meter stick to pass a given point in your frame. That would specify the length also. Do these two methods give different answers? The Lorentz transformation can be used to answer these questions. Consider a stick of length L_0, which is at rest in the primed frame. It extends from $x' = 0$ to $x' = L_0$. What are the position coordinates of the two ends in the unprimed frame at $t = 0$? According to the Lorentz transformation, equations at $t = 0$, $x' = \gamma x$. The stick extends from $x = 0$ to $x = L_0/\gamma$ at this time. In other words, the measured length of this stick is shorter than the original by a factor of gamma. It wouldn't have mattered if a different time had been chosen to make the measurement. The time was chosen to be zero for convenience.

It should be noted that by virtue of Einstein's interpretation, a meter stick that was at rest within the unprimed frame would also be considered shorter by a primed observer. The primed observer would also check the positions of the two ends simultaneously, but simultaneous to a primed observer would be at different times to an unprimed one. It doesn't make sense to think of the contraction as being a kind of compression. The contraction is just a way of telling us what a measurement would reveal.

Let's verify that the second way of measuring a length gives the same answer. Consider a stick of length L_0 in the prime frame that extends from $x' = -L_0$ to $x' = 0$. It is stationary in the primed frame. The origins of the two coordinate systems and the times coincide at $t = t' = 0$. What is the time when the "left" end of the stick reaches the origin as measured by the unprimed clock? Use of the Lorentz transformations leads to the following set of equations:

$$x' = \gamma(x - vt), \quad -L_0 = \gamma(0 - vt), \quad t = \frac{L_0 / \gamma}{v}.$$

In other words, the time it takes for the moving stick to pass a fixed point is the same as it would have been if the stick had shrunk by a factor of gamma. The two methods of measuring the length of a moving object agree with one another.

If Einstein's ideas confuse you, don't be embarrassed. Some of the best scientists in the world had trouble adapting to this new point of view. It may be helpful to ask yourself the following questions: "What is meant by the length of an object? Is it just a property of the object, or is it a property of the object plus that of the observer?" The theory gives no explanation as to how a wave can propagate through a vacuum. It just does. Nature need not comply with common sense. Einstein's theory is trusted because its predictions seem to be borne out in great detail.

You are encouraged to show that moving clocks run slower than stationary ones. In other words, consider two successive ticks of a clock that is stationary with respect to the primed frame and find the time duration between these ticks, as reckoned by someone on the unprimed frame. Note the word *reckoned*. It has been highlighted because you have to be careful to distinguish between what you observe, with your eyes and ears, and what actually happens, according to observers at the scene. Even in pre-relativistic physics, the observed time interval between hearing successive ticks of a clock depends on whether the clock is moving with respect to you. This difference is easily accounted for. For example, imagine you are in a stationary boat, and waves are lifting you up and down about every ten seconds. If you now turn on the boat's engine and travel slowly in the direction that the waves are traveling, you would increase the time interval between successive crests. The phenomenon is known as the Doppler effect. Note that the frequency may actually go up if the boat were to travel at a speed much faster than that of the waves. Einstein avoided the problem of timing distant events by imagining that there were an infinite set of clocks in each coordinate system. You should imagine that there are clocks embedded everywhere in the unprimed frame. These clocks are synchronized with one another in the same way that clocks on earth are synchronized. The same is true for the primed frame. It is possible to assign the time of an event by the clock located right at the event. The information would not be available to a distant observer immediately, but the record would be available for checking at some later time.

Consider the following question: "If a particle is moving parallel to the x-axis with velocity v' with respect to a primed observer, how fast is it traveling with respect to an unprimed observer?" The equation of motion for a particle in the primed frame is x' = v't'. The Lorentz transformation is used to express these variables in terms of x and t,

$$\gamma(x - vt) = x' = v't' = v'\gamma(t - vx/c^2)$$

$$x(1 + v'v/c^2) = t(v' + v)$$

$$x = \frac{v + v'}{1 + v'v/c^2} t.$$

In Newtonian physics, the denominator would be 1. In the special case where v' = c, such as the motion of a pulse of light moving along the x axis, the equation becomes

$$x = \frac{v+c}{1+v/c}t = c\frac{1+v/c}{1+v/c}t = ct.$$

Congratulate yourself if you anticipated this result. It is consistent with the assumption of the constancy of the velocity of light in all inertial frames.

Einstein wasn't the only scientist who was concerned about the constancy of the velocity of light. In addition to Lorentz, the Irish physicist George Francis Fitzgerald (1851–1901) had assumed that meter sticks contracted when in motion to account for the Michelson Morley results. The contraction was referred to as the Lorentz Fitzgerald contraction. The French mathematician Henri Poincare (1854–1912) was the first one to cast the Lorentz transformations in the symmetrical way that they were presented here. Einstein never mentioned any of these great scientists in his paper on relativity.

The Twin Paradox

The special theory of relativity involves certain strange ideas. When the world became aware of Einstein's theory, many people were disturbed by some of its predictions. One of these had to do with the biological process of aging. If moving clocks run slower than stationary ones but are otherwise normal in the frame in which they themselves are stationary, it follows that all biological processes, heart rate, breathing rate, etc., are slowed in the moving frame. If two identical twins were separated such that one of them stayed on the earth (a nearly inertial frame) and the other went on a long round trip, at a speed near the speed of light, there could be a large difference in their apparent ages when the traveling twin returned. This prediction of the theory was very upsetting to a great number of people. Many books were written critiquing relativity theory for such an absurd prediction. Nevertheless, most scientists believe it. There is a huge amount of evidence for it. An early piece of evidence had to do with the observation of mu mesons (now called muons). These are radioactive particles produced high in the atmosphere by incoming radiation. When produced in a laboratory, they have a very short lifetime. It is so short that very few, if any, should ever reach the surface of the earth when produced in the upper atmosphere. However, they are moving so rapidly through the atmosphere toward the earth that their clocks have slowed down, and they live long enough for scientists to observe them at ground level without any trouble.

Although no humans have experienced a velocity with respect to the earth sufficient to experience any measurable difference in aging, it would be wrong to say of the predicted difference "It is only a theory." The objectionable word in this expression is *only*. There is no difference between a law of physics and a physical theory. Nowadays, scientists are reluctant to call a proposition a law. However, although theories may be replaced by more accurate ones, the confidence level that scientists have in such predictions of relativity theory is enormous. It is a theory that has been tested over and over.

You might want to critique the twin paradox on the basis of symmetry. You might claim that from the traveling twin's point of view, it was the other twin that took the trip, and it is she who has a slower heart rate, etc. You would be overlooking one thing that destroys the symmetry of the situation. The twin who stayed on the earth remained on a single frame throughout the separation. The traveling one had to turn around. She transferred from one inertial frame to another. That transfer is enough to make all the difference. The arguments, having to do with how the trip appears from the point of view of the moving twin, are subtle. It will merely be shown that strange things happen when a transfer takes place from one inertial reference frame to another.

Suppose there is a star that happens to be stationary with respect to the earth. Its distance from the earth is millions of light years. Of course, when you observe it, you are seeing what was happening millions of years ago. Nevertheless, you could determine where it is right now as such and such. If you change inertial frames and approach this star at a fairly high speed, its distance from you will change. The change will be partly due to the fact that you are moving toward it, but also because of the Lorentz contraction of the space between you and the star. In other words, as determined by calculations in the new frame, the distance to the star has contracted by the factor gamma. This is a legitimate use of the Lorentz transformation. Even if gamma is very close to 1, differing by less than 1 percent, the star can be reckoned to be closer to you by perhaps hundreds of millions of miles. If it only took a matter of seconds to make the change in frames, it could be said that the star got closer to you by more than a million miles per second. This is a speed much faster than the speed of light. No such speed is attainable in an inertial frame according to relativistic theory. Although you don't observe the star to be moving, because the light from it hasn't reached you yet, the calculation is meaningful. It is possible for the speed of an object to exceed the speed of light in a non-inertial frame. You have to be careful when you make calculations based on such a frame. For this reason, you cannot rely on the imagined symmetry in the twin paradox. The twin who changes reference frames can't use arguments like the ones made by the twin who stays on one.

A pleasant aspect of the twin paradox is that the trip to a distant planet may be possible even if the planet is hundreds of light years away. You may question this. After all, if it takes light hundreds of years to get there, it will take people much longer than that. That statement is true. Nevertheless, the people making the trip need not age that much if they are traveling near the speed of light. Their biological clocks are not ticking fast, as seen from the earth. From their point of view, their biological processes seem normal. However, by their measuring apparatus, the star is much closer than hundreds of light years. Both the travelers and the ones left behind would agree on how much biological aging will take place. Please don't expect travel to such distant planets to take place in the near future. The energy needed to get a spaceship to such a velocity as to be meaningful is almost unimaginably large.

Mass and Energy

Einstein wrote a second paper on relativity in his miracle year of 1905. He considered two gamma rays of equal frequency being absorbed by a stationary object. The gamma rays are traveling in opposite directions so that they carry zero momentum. The momentum of the entire system is zero. He then considered what the situation would look like from a frame of reference that was moving in the direction of one of the gamma rays. Each gamma ray would be shifted in frequency and thereby each photon would have a different energy and momentum. In order to obtain energy and momentum conservation, he found it necessary to assume that the mass of the object was changed when its energy changed. The change in mass was directly proportional to the energy. It seemed that mass itself was a measure of the energy content of a body. Einstein came up with what is probably the most recognized equation of physics, namely,

$$E = mc^2,$$

an equation even more famous than F = ma. Einstein realized that because c is such an enormous velocity, the usual energy shifts from heating cause a minuscule mass change in a body. He suggested that the theory be checked by looking at the mass change associated with radioactivity. Looking at it from another point of view, it was recognized that a huge amount of energy could be released from a few grams of material if that energy could be converted to useful energy. A few years later when the possibility of a chain reaction involving uranium was recognized, it was this equation that made scientists aware of the enormous energies associated with nuclear fission and later with nuclear fusion. Einstein has been called the father of nuclear energy.

Four-Dimensional Space-Time

In classical physics, space and time are treated completely in-dependently. Time is considered a scalar while position is a vector. This last statement is not quite right. It is better to deal with the difference in position between two points in space. Let the Cartesian coordinates of one point be (x_1, y_1, z_1) and the other (x_2, y_2, z_2). These numbers depend on where the origin of the coordinate system is placed. However, the difference, $x_2 - x_1$, for example, doesn't depend on this choice. The change of position from point one to point two is a vector in three-dimensional space. The actual values of the three components of this vector depends on the orientation of the coordinate axes, but there are definite rules for transforming from one coordinate system to another. There is one scalar associated with a vector, namely, its magnitude. This is the distance between the two points, in the example just considered. The square of this distance can be written as

$$d^2 = (x_2 - x_1)^2 + (y_2 - y_1)^2 + (z_2 - z_1)^2.$$

Since this quantity is an invariant, the RHS would have the same form if primes had been used on each of these coordinates. The time interval between two events is also an invariant in classical physics.

Neither of these quantities are invariants in relativity theory. However, Hermann Minkowski (1864–1909), a German mathematician, pointed out in 1907 that the quantity

$$(x_2 - x_1)^2 + (y_2 - y_1)^2 + (z_2 - z_1)^2 - c^2(t_2 - t_1)^2$$

is an invariant. It is an interesting sidelight that Minkowski had been one of Einstein's math professors who hadn't thought much of Einstein's ability as a student. Einstein's miracle year must have seemed truly miraculous to him! Minkowski made the comment that it was no longer possible to think of space and time as separate entities. The concept of space-time had been born. Note that space and time are similar in some respects. The main difference is the factor c and the minus sign in the invariant. Note that the classical

238

invariant is positive while the relativistic one can be positive, negative, or zero. If it is positive, the separation between the events is said to be space-like. In that case, different observers can differ as to which event is earlier. If the invariant is negative, the separation is time-like. If one of these events is the emission of a light pulse and the second is the reception of this same pulse, the invariant is zero.

It was clear to Einstein and others that Newton's laws of motion had to be revised. For one thing, velocity and acceleration were no longer vectors. This can be seen from the fact that both quantities involve a rate of change with time. Since time is no longer a scalar, the associated quantities can no longer be thought of as transforming the way position vectors do. If this makes no sense to you, or if you think, "So what?" consider the following: Suppose experiments had been performed in some laboratory in which a coordinate system had been erected. The results of the experiment were consistent with Newton's second law. In other words, the following equations were satisfied:

$$F_x = ma_x, \ F_y = ma_y, \ F_z = ma_z.$$

Now consider the same experiment as viewed from a different frame of reference, using a different coordinate system. If the force transformed as a vector, and the acceleration transformed in some other fashion, the left side of the equations would no longer correspond to the ones on the right. The law would be valid only in one particular lab and have no applicability elsewhere. It certainly couldn't be considered to be a law of physics.

There is a way out of the fact that the velocity of a particle doesn't transform as a vector. Consider the way its average velocity is determined as it moves from one point to a nearby point. The change in position is divided by the elapsed time. The problem stems from the fact that the elapsed time differs from observer to observer. There is a way out of this difficulty. It involves imagining that the particle under investigation carries a clock with it. If all observers were to rely on that clock, they would agree on the time it takes for the particle to move from one point to the next. They would be dividing by the same quantity. In that way, the velocity would be a vector. This time interval is called the proper time. It is usually denoted by the Greek letter *tau*.

In order to make the momentum of a particle behave like a vector, it is necessary to redefine it. Thus, for example, the x-component of this vector is given by

$$p_x = m\frac{dx}{d\tau} = m\frac{dx}{dt}\frac{dt}{d\tau} = \frac{mv_x}{\sqrt{1-v^2/c^2}}.$$

In this equation, the letter v represents the velocity of a particle, not a reference frame. The factor in the denominator takes into account the fact that moving clocks run slower than stationary ones. There is still one problem. The changes in the position coordinates, dx, dy, dz, have been treated as components of a vector in three-dimensional space. In the four-dimensional world of space-time, a fourth component is needed. Vectors in space-time are called four-vectors. Instead of letter subscripts, numerical subscripts are usually adopted. The number 0 is used for the time, and 1, 2, and 3 for the spatial components.

$$p_0 = m\frac{d(ct)}{d\tau} = \frac{mc}{\sqrt{1-v^2/c^2}} = \gamma mc.$$

An invariant can be made out of out of four-vectors in a similar way to the one used by Minkowski with respect to location in space-time. It turns out to be more convenient to reverse the sign of this invariant in order to obtain a positive value for the answer. The invariant associated with momentum is, thus,

$$p_o^2 - p_1^2 - p_2^2 - p_3^2 = \frac{m^2c^2 - m^2v^2}{1-v^2/c^2} = m^2c^2\frac{1-v^2/c^2}{1-v^2/c^2} = m^2c^2.$$

This is clearly an invariant.

The question arises as to the interpretation of p_0. It depends on the mass and the speed of a particle. Could it be related to the kinetic energy? In the classical limit, v is much smaller than c. The quantity gamma could be well approximated by the first few terms of a power series . What is meant by a power series expansion for a function of x is an expression of the form

$$f(x) = c_0 + c_1 x + c_2 x^2 + \dots c_n x^n + \dots$$

In this expression, the constant coefficients are determined one at a time. The coefficient c_0 can be found by setting x to 0. The next coefficient can be found by first differentiating f(x) with respect to x and then setting the resultant function to 0. For the problem at hand, x is replaced by v^2/c^2. The power series expansion for gamma takes the following form:

$$\gamma = 1 + \frac{v^2}{2c^2} + \frac{3v^4}{8c^4} + \dots$$

No particle in the history of classical physics ever had a speed comparable to the speed of light. The third term on the RHS is more than millions of times smaller than the second term for any classical particle ever observed. It can be neglected. If only the first two terms on the RHS are retained, it is found that

$$p_0 c = mc^2 + \frac{1}{2}mv^2.$$

This formula has an easy interpretation. The first term on the RHS is interpreted as the rest energy, found earlier by Einstein. The next term is the classical value of the kinetic energy, the energy due to motion. Classical physics still works for ordinary slow-moving particles if a constant is added to the expression for the energy. It is only energy differences that matter. It is thus consistent to define the energy of a particle by the formula

$$E = p_0 c = \frac{mc^2}{\sqrt{1 - v^2/c^2}}.$$

The concept of momentum and energy have been joined together as part of a four-vector in a relativistic universe. Replacement of p_0 in an earlier equation by E/c leads to the following:

$$E^2 - p^2 c^2 = (mc^2)^2.$$

In a sense, the theory of relativity is very difficult in that it takes a while to get accustomed to such a new viewpoint. It also simplifies certain ideas. Things that were considered isolated from one another become connected and part of a bigger picture.

You should note that potential energy hasn't been included in this discussion. Although that concept was a good one in classical physics for dealing with a gravitational or electrostatic field, it is a dangerous concept to use in relativity. Although a field may be static from the point of view of one reference frame, it is not static as viewed from a frame in relative motion with respect to it. A full relativistic treatment that includes electrical potential energy, for example, would treat this quantity as part of a four-vector. It would then be necessary to include the spatial components of this four-vector. This is an advanced topic and will not be treated here.

Only some of the main ideas in relativity theory have been considered. It has been shown that classical physics suffered from the defect that time isn't a scalar, and velocity and acceleration aren't truly vectors. It turns out that the field quantities **E** and **B** of electromagnetic theory are not vectors either. They are part of a quantity called a tensor. Einstein was able to show how these quantities can be calculated in any reference frame if they are known in one. He single-handedly revised Newton's equations into a form suitable for dealing with particles of any energy. Today there are accelerators that produce particles that travel at speeds nearly as great as c. The design of these accelerators rely on the accuracy of relativistic laws. The theory has been severely tested. It is hugely successful.

The General Theory of Relativity

The special theory of relativity wasn't enough for Einstein. His mind was forever active. He wondered about a number of things. He knew that his theory of relativity was restricted to certain special frames of reference. He also wondered about certain coincidences. He thought about Galileo's experiments and Newton's law of gravitation. Why should the force and the inertia conspire to make all masses accelerate in the same way? He thought about freely falling elevators in which it was impossible (almost) to detect a gravitational field. Passengers would be falling at the same rate as the elevator and need not make contact with the floor. He also imagined a similar kind of elevator in a region in which there was no gravity. If a rope were attached to it and a force applied to it, such an elevator could be made to accelerate. This would look like a gravitational force field to the people inside. The idea that a gravitational field could be simulated by an accelerated frame of reference is called the principle of equivalence. The difference would only be noted if the elevator were big enough so that tidal forces would be noticed. There would be no tidal forces if the field were produced by the pull of a rope.

Could it be that real gravity and accelerated frames shared something in common? It was the beginning of an idea. He imagined a light beam coming though a window of an elevator as it was being accelerated. In the time it took for the light to cross the elevator, the elevator would have changed its velocity. The beam would be bent. Would a light beam be bent by a gravitational field? He wondered whether he could find a set of equations or laws that would be valid in any frame of reference. He had the special theory of relativity to guide him. He was dealing with a complex system in which masses played an active role in determining the behavior of systems in their vicinity as well as a passive role in that their behavior was controlled by other masses in the environment. It was as though the four-dimensional world of space-time were curved. The curvature was caused by objects and in turn affected the motion of other objects. Some authors explain this as though all bodies are located on a trampoline and deform it in such a way as to cause other objects on the trampoline to be affected.

The idea of a freely falling elevator made Einstein realize that even in a strong gravitational field, it was possible to find a frame of reference in which the theory of special relativity was a very good approximation, at least locally. The problems he faced made him realize that he needed the help of mathematicians. He studied the mathematics of non-Euclidean geometry and the theory of tensors. These were very difficult subjects, familiar mainly to mathematicians. He spent the years between 1911 and 1915 doing little else but working on the theory. This was work in which he was essentially alone, although he had revealed his progress to others in talks and papers. As he was coming close to finishing this work, he had confided to his friend, the great mathematician David Hilbert (1862–1943), the problems he was facing. Hilbert started working on the problem, and there is a dispute as to whether Hilbert or Einstein completed the theory first. Hilbert acknowledged that Einstein deserves all the credit for the theory, and almost everyone thinks it is the greatest theoretical accomplishment of all time. It is in a class by itself.

The only successful prediction of the theory at the time of completion had to do with the orbit of the planet Mercury. As shown earlier in this book, Newton's theory predicts an elliptical orbit for all the planets. According to Newton, perihelion, the point of closest approach to the sun, is a fixed point in the inertial frame of the solar system. When astronomers included the effects of all the planets in the solar system, it was found that the theory predicted that the perihelion would gradually rotate about the sun. This rotation was miniscule, and during the course of a century, the rotation would be less than 1°. However, astronomical precision is so great that this rotation was easily measurable. The only problem was that there was a discrepancy between theory and observation. The discrepancy was a mere 43 seconds of angle per century. Note that a second is one-sixtieth of a minute, and a minute is one-sixtieth of a degree. When Einstein used his newly postulated theory to calculate the precession of the perihelion he got 43 seconds of arc per century. He claimed it to be the greatest moment of his life.

At the time of Einstein's accomplishment, the world was immersed in a world war. There was little note of his work outside of the academic community. There was another prediction of the theory that was important but couldn't be tested. The theory predicted that light from a star that passed near the surface of the sun would be bent. The bending that his new theory predicted was twice the amount that a naive calculation, based on the equivalence principle, would have predicted. The equivalence principle is the idea that started Einstein on his search for a complete theory. When he first latched onto this thought, he probably had no idea of the twists and tangles that the complete theory would involve.

In order to test the theory, it was necessary to await the occurrence of a total eclipse. The opportunity came in 1919, shortly after World War I came to a close. On May 19 of that year, Sir Arthur Eddington (1882–1944), a famous British astronomer, led a group of astronomers to an island off Africa to test Einstein's prediction. The test proved successful, and Einstein became a celebrity overnight. Although there weren't many phenomena that could be used to test the theory in the early part of the twentieth century, it is now a well-established theory. One of the gadgets in common use today, namely, the navigation system called GPS (global positioning system) depends very sensitively on the theory. Clocks in orbit in a gravitational field have to be adjusted to be synchronized with those on the surface of the earth. It is necessary to make corrections consistent with Einstein's theory in order for these gadgets to work properly. If these adjustments are not made, the calculated positions of the devices would be off by hundreds of miles in less than a day.

The theory also predicted the existence of black holes. These are astronomical objects that are so massive and so dense that the gravitational force causes them to collapse. As far as we now know, there is no limit to how dense they can get. Although no one believes they can collapse to a point, our present theories are not good enough to know how small they can get. There are many things that are not known about black holes, but there is little doubt that they exist. At the center of our galaxy, there is one that has the mass of about 4 million solar masses. They are called black because the gravitational field is so strong that nothing can escape from them, not even light. They are usually examined by the way they affect stars in their neighborhoods. There is often a lot of radiation in their neighborhoods as objects swirl inward as they are being caught. Stephen Hawking has studied them theoretically and has shown how they are able to radiate away some energy, but the mechanism is subtle and will not be discussed here. It does not seem to radiate as a block body and the entropy of such a system is a subject of much discussion.

When Einstein first published his theory, he thought, along with many other scientists, that the universe didn't change much with time. For all the world knew, it might have existed forever, stars dying and being born all throughout history. At the time of the publication of Einstein's theory, the only thing that was known of the universe was our galaxy, the Milky Way. It took several years for Einstein to realize that his theory predicted that the universe would have to be expanding or collapsing. At the time he presented the theory, there were few solutions to his equations. An important solution to his equations was obtained by Karl Schwarzschild (1873–1916), a German scientist. Scwarzchild died a little over a year after the theory was published, and before it was confirmed by Eddington. What is even more remarkable is that he obtained a solution while he was on active duty in the German army during World War I. His solution involved the determination of the gravitational field around a spherical mass. His calculations indicated that a star would collapse under its own gravitational field if it were sufficiently dense. He essentially predicted the existence of black holes. He and Einstein also realized that the theory predicted either the expansion or the collapse of the universe. Although Einstein didn't believe black holes actually existed, he became aware that his theory couldn't account for a steady state universe. He modified the theory slightly to include a repulsive term that would keep the universe from collapsing. He made this "correction" about two years after publication of the original theory. Later, when the famous American astronomer Edwin Hubbell (1889–1953) made his discoveries of other galaxies, all that changed.

Hubbell's discoveries gave evidence that the universe is expanding. This is a tricky subject, so it will be necessary to make use of an analogy to make this point clear. What Hubbell actually discovered is that galaxies are receding from us at a rate that is proportional to their distance. Astronomers have learned not to think of the earth as being at the center of anything. The way to explain Hubbell's observation in our three-dimensional world (four if you think in terms of space-time) is to consider the two-dimensional world corresponding to the surface of a spherical balloon. Imagine that this surface has small insects on the surface that can move around. The insects correspond to stars in our three-dimensional world. They may live in colonies that correspond to galaxies, but there are large regions of space between these colonies. All that these insects or things living on them can perceive are on the surface of this two-dimensional surface. However, outsiders can see that the balloon is being blown up. It is expanding. The colonies are getting farther away from one another. It isn't easy to notice the distance between objects changing if they are close. Distant objects are moving away faster. Each colony can think of itself at the center of a region that is moving away from it. Except there is no center, as far as the two-dimensional surface is concerned. Although it is hard to imagine, it is possible to say that our universe might be finite and still have no boundaries. Think again of that balloon. From the point of view of a creature that can only see the surface, it is finite, but it has no boundary. If our universe is finite, we cannot think of it as expanding into a vacuum. It can be finite, perhaps, without any boundary. These mind-boggling ideas are thrust upon us by observations that have no simple explanation. Our ancestors must have had a similar problem with absorbing the fact that the earth isn't flat.

Hubbell's discovery led up to the Big Bang Theory. The theory essentially postulates that the whole universe was at one time smaller than the nucleus of an atom. There are detailed conclusions about how this tiny entity inflated enormously in some very early moments. The early universe was tremendously hot, and there were all sorts of radiation throughout. It is claimed that this radiation still exists, although it corresponds to a temperature of only a few degrees above absolute zero (-273^0 C). This radiation has been studied in great detail. By looking at what corresponds to temperature fluctuations of a millionth of a degree or so, scientists can detect patterns in this radiation that correspond to the distribution of matter in the universe.

When Einstein learned of the expanding universe, it became apparent to him that he didn't need to put in a correction term to his theory. Einstein claimed that this correction term was his worst mistake. He should have been able to postulate a changing universe before it was observed. However, discoveries that were made, not too long ago, make Einstein's apparent blunder look as though it wasn't a blunder after all. An exploding universe is expected to slow up in much the same way as a ball thrown upward does. Recent discoveries indicate that the expansion is actually speeding up. Cosmologists attribute this to dark energy. This is essentially equivalent to Einstein's correction term.

There are still many mysteries left. The motion of many stars cannot be accounted for without assuming that there is some matter in the universe that we cannot see. The only evidence for this so-called dark matter is through its gravitational influence on visible matter. The evidence actually points to the fact that dark matter accounts for more than three-quarters of all the mass in the universe. Although some scientists claim that there are theories that predict the existence of matter that interacts only weakly with light and other radiation, there is still much mystery about why such particles haven't been found.

Einstein stands out in the minds of today's physicists as Newton did in the classical period. They both changed the way the world looks at things. Newton did away with the idea that astronomical objects were guided by angels. He made earthly phenomena explainable in terms of rules or laws that applied to all objects. Einstein embraced ideas that seemed completely foreign. Atoms really did exist. We now know how big they are. Light doesn't require a medium to propagate like a wave. Not only is it a wave but it has attributes like that of a particle. A theory must be found that allows for both these attributes of light. Maxwell's equations can't be relied upon as being the ultimate description of light. The notion of absolute time is a myth. What is the meaning of the length of a moving object? Scientists not only learned how to absorb all these amazing new ideas, they also learned to be bolder in the way they thought about things. The quantum idea, pounced on by Einstein as being more than a mathematical device, led to bold new ideas that even Einstein couldn't accept.

Quantum Theory

Unlike the theories discussed in most of this book, the story of the quantum evolved gradually. There are a huge number of names associated with it. In order to do the topic justice, a book devoted to that one subject alone is required. Only the most significant developments will be highlighted. In 1907, Einstein applied the ideas of Planck to the explanation of why the specific heat of solids approached 0 as the temperature approached 0. Basically the idea was that a solid behaved as though it were made up of a collection of oscillators that were vibrating randomly when subjected to being heated. Classical theory assumed an oscillator could have any energy and that the energy would be proportional to the temperature. Thus, it would take as much heat energy to raise a sample's temperature one degree from 50 to 51 as it would from 1 to 2. In other words, the specific heat should be constant if this model is correct. However, the specific heat dropped toward 0 at very low temperature. Einstein took this as good evidence for the Planck model. According to this model, an oscillator required a reasonable chunk of energy before getting out of its lowest energy state. It just couldn't take the small bits of energy available at low temperatures. Einstein saw this as the main idea. However, he used an oversimplified model in which only one frequency was assumed for the oscillators. It got to the heart of the matter. Later, Peter Debye (1884–1966), a famous chemist, made a more accurate model that fit the experimental data beautifully. Instead of picturing oscillators that had only one frequency, he used classical elastic theory to envision a huge collection of different modes of oscillation of a solid body.

In 1912 the great Australian physicist Ernest Rutherford (1871–1937) discovered the atomic nucleus. By using a beam of alpha particles emitted from a sample of uranium he discovered that more than 99 percent of the mass of an atom was contained in a positively charged particle at is core. It was approximately a thousandth the diameter of the atom. That brought up a huge question. If electrons were in orbit around such a nucleus, they would emit radiation and lose energy according to classical reasoning. What kept them in orbit?

Just one year after Rutherford's discovery, Niels Bohr (1885–1962) came up with a response to this question that involved some bold assumptions. He restricted his attention to the hydrogen atom, consisting of one electron in orbit about a single particle, the proton. He assumed that the angular momentum of the particles was quantized. In other words, the angular momentum L was given by the formula

$$L = n\frac{h}{2\pi}. \quad n = 1, 2, 3...., $$

where h is Planck's constant. The factor of 2π was needed to explain Planck's result when applied to three-dimensional oscillators. With this assumption, only certain orbits and certain energies were possible. Another assumption was that the atom didn't radiate when the electron was in one of these orbits. Bohr assumed that the atom could only absorb light when a photon had the precise energy to cause a transition between two levels. It was the type of relationship that Einstein had used in the photoelectric effect. In other words, the product of Planck's constant with the frequency of the light had to correspond to the difference in energy between the level in which the atom found itself and a higher one. Typically, the atom would be in the lowest level. If the atom was excited to a higher level by some stimulus, it could emit radiation of a definite frequency when it made a transition to a lower level. Amazingly, these assumptions corresponded to the spectrum that was observed when atomic hydrogen was exposed to radiation.

Bohr's successful description of the hydrogen atom had a huge impact on spectroscopy. For example, if you could determine the frequencies present in a neon tube, you would have evidence for the values of the possible energies of the neon atom. It became clear that energy came in chunks. It no longer was possible to ignore the quantum hypothesis.

Einstein made a contribution to the theory of the absorption and emission of photons by atoms in 1916. It was known that an atom could emit light by spontaneously dropping from a high energy level, E_H, to one of lower energy, E_L. This produced light of a particular frequency. It was also known that if an electron was in level E_L, that light of this frequency could be absorbed and cause a transition to E_H. The first process is called spontaneous emission, while the second process is called induced absorption. By considering an atom in thermodynamic equilibrium with black body radiation, Einstein was able to show that there was another process called induced emission. In other words, light of a definite frequency that was incident on an atom could contribute to the emission of that frequency, as well as to its absorption. There were certain constants involved in these processes that determined the probabilities of the various processes for a given level of illumination. Einstein provided formulas for these. The idea of induced emission later led to the invention of the laser by the american physicist Charles Townes (1915–2015). Townes, a Nobel Prize winner, was a pioneer in a new field called quantum optics, in which much activity is still taking place.

After 1913, the struggle to make sense of atoms that were more complicated than hydrogen was just that, a struggle. It wasn't easy to deal with more than one electron in orbit about a nucleus. Progress in understanding atoms came slowly. However, Einstein's theory of photons got an important boost in 1923. In that year, Arthur Compton (1892–1962) discovered that the scattering of X-rays by electrons in a metallic solid could be explained by assuming that light consisted of photons, with energy $h\nu$, and momentum $h\nu/c$. This is just the right relationship between energy and momentum given by relativity for a particle without mass. Einstein's postulate about photons was now accepted by the majority of scientists.

In 1924, Einstein was sent a paper by an Indian physicist, Sanyendra Nath Bose (1894–1974), who was able to reproduce Planck's results for black body radiation by applying statistical ideas to a gas made up of photons. Recall that the radiation inside an enclosure at a given temperature has the same distribution of frequencies as black body radiation. Einstein liked Bose's ideas so much that he translated the paper into German and had it published in a prestigious German periodical. He also published a few papers in which he applied the same statistical idea to atoms. The basic difference between quantum statistics and classical statistics is that identical particles could not be distinguished from one another, even in principle. The idea can be illustrated by using as an example a pair of dice. In determining the odds of rolling a six (the probability that on a given throw of the dice the sum of the two numbers on top will add up to six), it is claimed that there are thirty-six possibilities and that there are five ways of obtaining a six. These ways can be enumerated by writing down possible throws as (1, 5), (2, 4), (3, 3), (4, 2), (5, 1). These possibilities distinguish (1, 5) as being different from (5, 1), and (2,4) as being different from (4, 2). Einstein and Bose claimed that in quantum theory there are only three possibilities. It is not possible to distinguish between (2, 4) and (4, 2). In quantum theory there is only one possibility, in which one of the dice comes up 2 and another comes up 4. Of course, these ideas can't be applied to real dice. They can be observed as being at distinctly different positions. There is no question which is which. If they were quantum objects, it would not be possible to treat them as having separate identities. All that could be said is that one of them came up 4 and another came up 2.

It turns out that the Einstein-Bose theory cannot be applied to all types of particles. It will be seen that there is a property associated with many atomic particles called spin. In other words, they behave as if they have an intrinsic angular momentum. These are quantized in a way that is reminiscent of the Bohr theory, in which the orbital angular momentum is also quantized . If the intrinsic angular momentum of the particles is 0, or an integer multiple of $h/2\pi$, they obey Einstein-Bose statistics. They are called bosons. One of the things that Einstein and Bose discovered is that at low temperatures there is a tendency for bosons to collapse into the lowest energy state. This is called Einstein-Bose condensation. This phenomenon was not observed until recently. It is an interesting state of matter.

Bohr's ideas with respect to energy levels led a brash young Viennese theorist, Wolfgang Pauli (1900–1958), to explain the periodic table of the elements. He imagined an atom that contained many electrons having orbits like those in the hydrogen model. Pauli assumed that only two electrons could be in any given orbit at a time. The factor of two was needed in order for the calculated orbital picture to correspond to the periodic table. He assumed that the electron was not merely a point particle. It behaved as a body that could be spinning in one of two directions in a given orbit. Pauli's exclusion principle essentially allows only one electron to be in a given "state" at one time. A state consists of an electron of a particular spin being in a particular orbit. There are actually many orbits that have the same energy in the Bohr model. These orbits can be regarded as constituting a shell of charge surrounding the nucleus. In order for the calculated numbers to fit the periodic table, Pauli's idea was that a system in which a set of shells was full would be chemically inactive. It would correspond to the noble gases. A single electron in orbit outside of a closed shell could give up its electron easily. It would act like an alkali metal. He put forth these ideas in 1925.

It should be noted that the electron is not a boson. Remember, there can be any number of bosons in a given quantum state, while only one electron can occupy a given quantum state. It doesn't satisfy the requirements specified earlier. In units of $h/2\pi$, the electron's angular momentum is $1/2$. Pauli proved that any particle that has a half-integer spin, like that of the electron, satisfies the exclusion principle. It turns out that protons and neutrons also have such a spin. Such particles satisfy a different kind of statistics than the one associated with bosons. The statistics was worked out by Enrico Fermi, and spin half particles are often called fermions.

The spin of a system can make a huge difference in its properties. For example, consider atoms of helium. Each atom of helium consists of a nucleus that contains two protons and some neutrons. The more common atom is one that contains two neutrons. The other one consists of a nucleus with just one neutron. They are said to be different isotopes of helium. The chemical properties of most atoms are determined by the number of protons (or the number of electrons, since they are the same). In a simple-minded picture, the chemistry is the result of the orbiting electrons in the outermost shell being available to move from one atom to another. Isotopes of an element don't matter much in this simple picture.

The common isotope of helium is called helium four because it has four nuclear particles. The main difference between helium four and helium three has to do with the fact that the former is a boson and the latter is a fermion. When a combination of particles exist in a system, its spin is a combination of the spins of the ingredients. An even number of fermions combines to form a particle with an integer spin. Helium four consists of an even number of electrons, an even number of protons, and an even number of neutrons. It has an integer spin and is therefore a boson. The fact that helium three has one neutron leaves it with a half-integer spin. It is therefore a fermion. When cooled, helium four liquefies and at a temperature about 2° Celsius above absolute zero it becomes a superfluid. It has seemingly miraculous properties in that part of the fluid loses all traces of friction. It has zero entropy. It can climb walls and flow through pores. The fact that the phenomenon is related to spin indicates that it is a type of Bose-Einstein condensation that is the primary cause of this behavior. It is true that helium three also becomes superfluid, but this phenomenon occurs at a much lower temperature. Theorists explain the superfluidity of helium three as being a consequence of pairs of atoms correlating their motions in such a way that each pair behaves like a boson.

Wave Particle Duality

In 1924, a PhD thesis submitted by Louis De Broglie was brought to the attention of Einstein. De Broglie had thought, "If light has particle-like properties, wouldn't it be possible that particles, such as electrons, have wave-like properties?" He suggested that the wavelength associated with a beam of electrons of a given velocity should equal Planck's constant divided by the momentum

of a single electron. The thesis committee wanted Einstein's opin-
ion before accepting these ideas. Einstein thought the analysis was
good and gave his hearty approval. De Broglie would get the Nobel
Prize a few years later.

What made De Broglie associate the wavelength of a sinusoidal
wave with the momentum of a particle? To understand this, it helps
to understand the relationship of the wavelength of a light wave to
the momentum of a photon. The relationship between energy and
momentum of a particle was given by Einstein's theory as

$$E^2 - p^2c^2 = m^2c^4.$$

This can be applied to a photon, a particle with no mass. In that
case, the energy is just the product of the momentum with the velocity of light. This relationship can be written in the following way:

$$E / p = c.$$

If the energy is expressed in terms of the frequency of the wave, the
result is

$$hv / p = v\frac{h}{p} = c.$$

It is a well-known property of traveling waves of a given frequency
that the product of the frequency with the wavelength is the wave
velocity. This led De Broglie to his assumption regarding the relationship between the momentum of a particle and the wavelength
of the associated wave, namely,

$$\lambda = h / p.$$

Two American physicists, Clinton Davisson(1881–1958) and
Lester Germer (1896–1971), confirmed De Broglie's hypothesis by
experiments carried out between the years 1923 and 1927. They
found that a beam of electrons of a given momentum is scattered
by a crystalline materials in a way that is similar to the diffraction
of a beam of X-rays. The relationship between the momentum and
the wavelength was just that of the theory. The time was ripe for
revolutionary ideas. Two German theoretical physicists, Werner
Heisenberg (1901–1976) and Ernest Schrödinger (1887–1961),
had different approaches to the wave particle duality. Paul Dirac

(1902–1984), a British theorist, later showed the two approaches to be equivalent. Schrödinger's theory is easier to understand.

Schrödinger sought an equation similar to the wave equation used by Maxwell. There were certain requirements that this equation had to fulfill. The ideas of Bohr and those of De Broglie made it seem likely that an atom and a violin string had something in common. For example, a violin string of a given length and tension could vibrate only with certain possible frequencies. The electron in the hydrogen atom could only exist at certain energies. It was possible that Bohr's results could be obtained by making use of the wavelike nature of the electron. The frequencies of the particle waves could correspond to the energy levels of the atom. Schrödinger hoped his wave equation would give the correct frequencies for the hydrogen atom. He also wanted his theory to describe a free electron. He used classical physics to note that the relationship between the momentum and the energy of a free particle is given by the formula

$$E = \frac{mv^2}{2} = \frac{(mv)^2}{2m} = \frac{p^2}{2m}.$$

He wanted the frequency of his waves to correspond to the energy when multiplied by Planck's constant. He wanted the momentum to be inversely related to the wavelength such that

$$E = hv, \quad p = h/\lambda.,$$

These requirements gives a relationship between frequency and wavelength as follows

$$hv = \frac{h^2}{2m\lambda^2}.$$

This result was enough to convince him that the usual wave equation wouldn't work. The wave equation predicts that the product of the frequency with the wavelength is a constant. For electromagnetic waves, this constant is the speed of light. Schrödinger needed waves that had variable speeds.

In order to understand how Schrödinger arrived at his equation, it helps to think about a vibrating string. The solution involves the product of a sinusoidal function of time with a sinusoidal function of position. The wave equation involved a second derivative with respect to time and a second derivative with respect to position. When these operations were applied to the assumed solution, the functional form was unchanged. In other words, the same mathematical function was returned, except that it was multiplied by some constant. When the second derivative of time was applied, the constant turned out to be proportional to the square of the frequency of vibration. When the second derivative with respect to position was applied, the constant was related to the wavelength of the associated travelling wave. Actually, it was proportional to the square of the reciprocal of the wavelength. The end result of applying the operations of the wave equation was that a relationship was obtained between the frequency and the wavelength of these waves. The wave function factored out of the equation.

Schrödinger might have been thinking of waves on strings when he came up with his equation. The idea that second derivatives with respect to space yield a constant that's inversely proportional to the square of the wavelength was perfect. This can be interpreted as a constant that's proportional to the square of the momentum. For a free particle subject to no forces, the energy is proportional to the square of the momentum. Schrödinger's problem had to do with the fact that he needed an equation that gave the frequency, not the square of the frequency. The reason had to do with the fact that it was the energy that was needed, not the square of the energy. If the second derivative yields the square of the frequency, wouldn't the first derivative yield the frequency? That would be great, except... The problem had to do with the fact that the first derivative of the sine yields the cosine. The first derivative of the cosine yields the sine (except for a negative sign). He wanted the operations of his equation to give the same function back except for constant factors.

The crucial thought came that by combining the cosine and the sine in some way, the first derivative with respect to time would give the same function back. Suppose he tried the following function for the time-dependent part of the wave,

$$\psi = \cos(\omega t) + i\sin(\omega t),$$

where i is some adjustable constant. Note that the letter i has been chosen with malice aforethought. This is the symbol used in mathematics for the square root of negative one. It is the quantity that is needed for the purpose. as will be shown. Differentiation with respect to time yields

$$\frac{d\psi}{dt} = -\omega[\sin(\omega t) - i\cos(\omega t)]$$

$$= i\omega[\cos(\omega t) - \frac{\sin(\omega t)}{i}].$$

This equation will be a multiple of psi if

$$i = \sqrt{-1}.$$

In that case,

$$\frac{d\psi}{dt} = i\omega[\cos(\omega t) + i\sin(\omega t)] = i\omega\psi.$$

Consider the following operation, O,

$$O = -i\hbar\frac{\partial}{\partial t}.$$

The slashed quantity is aitch bar or aitch slash. It is Planck's constant divided by 2π. When applied to the time-dependent solution, this operation yields

$$-i\hbar\frac{\partial\psi}{\partial t} = -i^2\hbar\omega\psi = \frac{\hbar\omega}{2\pi}\psi = h\nu\psi = E\psi.$$

Actually, Schrödinger used a negative sign for omega in his time-dependent solution. So that the energy operator in his equation has a positive sign. The actual equation for the case in which an electron is in some conservative force field is

$$-\frac{\hbar^2}{2m}\nabla^2\psi + V\psi = i\hbar\frac{\partial\psi}{\partial t}.$$

In this equation, several "shorthands" have been used, namely,

$$\hbar = \frac{h}{2\pi}, \quad \nabla^2\psi = \frac{\partial^2\psi}{\partial^2 x} + \frac{\partial^2\psi}{\partial^2 y} + \frac{\partial^2\psi}{\partial^2 z}, \quad i = \sqrt{-1}.$$

The Greek letter *psi* denotes the electron wave function. The letter *V* specifies the potential energy function for an electron in the given force field. For the case of the hydrogen atom, it is the potential energy of an electron in the electric field of a proton. The letter, m, denotes the electron's mass. The reason a special symbol called aitch bar has been introduced is because it occurs so frequently. For example, Bohr had assumed that the angular momentum of an electron in the hydrogen atom was a multiple of this constant.

There are three operations performed on the wave function in the Schrödinger equation. The one on the right-hand side has been called the energy operation. The first term on the LHS is called the kinetic energy operation. Note its resemblance to classical expression for the kinetic energy, namely $p^2/2m$. The remaining operation is just the multiplication operation in which the product of the wave function with the potential energy function is carried out. This equation gives a meaningful solution when V=0, and it agrees with Bohr's energy levels for the hydrogen atom.

Heisenberg's formulation was different, but he also represented classical quantities by mathematical operations. Whereas Schrödinger had used the operation of differentiation with respect to x to correspond to the x-component of momentum, Heisenberg had used an operation involving a mathematical operation known as matrix multiplication. His method was known as matrix mechanics. He was able to conclude that certain measurements had to disturb the measurement of other quantities. He became famous for the uncertainty principle. He claimed it is impossible to know, for example, the position and the momentum of a particle at the same time. The measurement of one of these quantities has to interfere with the measurement of the other. He gave experimental examples. If you wanted to measure the position of an atomic particle by shining light on it, the photons would carry some momentum and knock the particle away. The more precise the measurement, the shorter the wavelength that was required. Such light would give more momentum to the particle. There seemed to be no way out.

Not only did Schrödinger's equation work for the hydrogen atom. it worked for the harmonic oscillator also. By this is meant that the spacing between successive levels is the same as that which was obtained by Planck. The only difference is that the lowest possible energy wasn't zero. Its value was half the energy difference between levels. This result is consistent with Heisenberg's uncertainty principle. If the oscillator had zero energy, the location and momentum of the particle would be known. It would have zero momentum, and its location would be at its classical equilibrium point.

This was the beginning of a new era in physics. One year later, Paul Dirac tried to combine Schrödinger's ideas with Einstein's relativity theory. The relationship between momentum and energy is different in relativistic physics. You may recall that the formula is

$$E^2 - p^2c^2 = (mc^2)^2.$$

Dirac knew that Schrödinger had good reason to use an equation that only involved the first derivative with respect to time. However, use of Schrödinger's idea, in which physical quantities are replaced by mathematical operations, presented a problem. Because the relationship involved the square of E, the equation seemed to require a second derivative with respect to time. There were good reasons to avoid this possibility. The alternative was worse. If the relativistic equation were solved for E, Dirac's wave equation would involve a square root. Neither alternative was acceptable to Dirac. He was able to get involved with the first derivative by what seems like an ingenious mathematical trick. He knew the electron didn't actually behave like a particle. There was evidence it had spin. This had already been used by Pauli in his exclusion principle. By making use of the fact that the electron had this extra attribute, he was able to get a set of equations that involved only the first powers of the energy and the momentum. It was consistent with the relativistic view that treated energy and momentum as parts of a four-vector.

Dirac's theory was hugely successful. Not only did it account for fine details in the spectrum of the hydrogen atom, it was the first theory that combined the special theory of relativity with quantum theory. It was also the first theory that explained the electron's spin in a natural way. It is most famous for the fact that it predicted the existence of antimatter. These are particles that can combine with particles of matter in such way so that both particles are destroyed. The huge amount of energy released emerges as radiation. Almost every theoretical physicist marvels at the ingenuity of Dirac and the beauty of the Dirac equation.

For a while there seemed to be a problem with the equation. It predicted negative energy states, not just a few but an infinite number of them. This was not only true for the hydrogen atom, but it applied to a free particle as well. There was no lowest energy level for the system to be in. The only way out seemed artificial. It was necessary to assume that all these negative energy states were filled and that Pauli's exclusion principle could be used to show that the lowest available state were the positive energy ones. It sounds a bit ridiculous in that the solution implied an infinite negative charge. This charge was waved away as though there was an infinite positive charge to compensate for it. This seeming failure of the theory turned out to be its strongest success. If the theory had negative energy states that were occupied with electrons, it might be possible to cause a transition from one of these occupied states to a positive energy state that was unoccupied. When the theory was applied to a vacuum, it predicted what is now known as pair creation. A photon with sufficient energy could cause a transition so that an electron would occupy a positive energy state, while the hole left in the sea of occupied states would act like a positively charged particle with the same mass. There was no such positive particle at the time. Dirac thought the hole could possibly be a proton, but the mass of the proton was much too large. The positron was discovered later, in 1933.

It was soon realized that the Dirac equation was a many particle theory. As such, it should allow for possible interactions between the particles that were created. It was also recognized that relativity didn't allow for action-at-a-distance forces so that these interactions had to be mediated by particles, such as photons. A new type of theory called quantum field theory was born, which adopted many of the features of Dirac theory without the necessity of a filled sea of negative charge. This theory is much too sophisticated to be treated here.

The non-relativistic theory of Schrödinger was generalized to deal with systems of many electrons. These ideas were used by chemists and physicists to explore the properties of atoms, molecules, and solids. Shortcuts were used to avoid the difficulty of considering the detailed nature of the interactions between the large number of electrons in such systems. In a way, the approximations were similar to the ones used by astronomers in dealing with the solar system. As a first approximation, you treat the interaction of a planet with the sun. Later, you make corrections because of the influence of other planets. The situation in chemical systems is a bit more complicated than that, but the exclusion principle turns out to be helpful, if the main interest is in the chemical properties of such systems. It turns out to be possible to focus attention on just those electrons that are called valence electrons. These are the electrons that are outside of the closed shells associated with tightly bound orbits.

Quantum theory turned out to be very successful in explaining why certain crystalline solids are metallic, others are insulators, and still other materials, known as semiconductors, have intermediate properties. Instead of discrete energy levels, solids have energy bands. In other words, there are ranges of energies in which electrons can exist, and forbidden ranges in between. In an insulator, these bands are all filled. It is similar to an atom of a noble gas in which all the shells are occupied. In metals the bands are only partially occupied. The electrons are free to be excited by taking a little bit of energy and moving about through the whole crystal. Semiconductors are like insulators except that only a little energy is needed to excite an electron over the forbidden region of energy. By raising the temperature of a semiconductor, electrons could be lifted from the valence band to the conduction band. They become better conductors as the temperature is raised. This is the opposite of what happens in metals.

There was a great deal of success explaining such mysterious phenomena as the superconductivity of lead and mercury at very low temperatures by means of quantum theory. Not only does all electrical resistivity disappear in these systems, they have strange magnetic properties as well. The explanation of these phenomena is complicated, involving indirect interactions between electrons. In a way, an electron can move through a region and cause the nearby atoms to react. The reaction, in turn, causes other electrons in the neighborhood to be affected. It is amazing that quantum theory was better suited for dealing with such phenomena than classical theory ever was.

There are still many unexplained phenomena, at least at the time of this writing. Some high-temperature superconductivity has not yielded to theoretical analysis, for example. However, quantum theory has made remarkable progress and accounts for much of the behavior of physical systems. The understanding of quantum optics has led to the development of lasers and all sorts of sophisticated apparatus in which lasers are used. Semiconductors are fairly well understood, in that the effect of miniscule amounts of impurities can dramatically affect their electrical properties. The computer chip would not exist if it weren't for quantum theoretical analyses made in the first half of the twentieth century. It is hard to think of any modern device that doesn't owe its existence to quantum theory.

Gravitation and Quantum Theory

There are certain features of the theory of gravity and those of quantum theory that lead to interesting questions. You have seen that light has properties that are wave-like and properties that are particle-like. These ideas, together with the first law of thermodynamics, lead to a strange conclusion, namely that clocks at the bottom of a tall building run slower than those at the top. In order to show this, consider a photon coming from the sun that is absorbed by an object at the top of some building. The object now has more energy so it is more massive, by virtue of Einstein's formula $E = mc^2$. Its potential energy in the gravitational field of the earth has also gone up. It is heavier after the absorption, and it has more potential energy than it would have at the bottom of the building. If the same photon had been absorbed on the ground floor, the object would have gained some energy but not as much. Energy conservation seems to be violated. In order to preserve the first law of thermodynamics the frequency of the photon must have gone up if it had fallen past the roof to the ground. However, this brings up another question. If a light wave of a single frequency is being considered, it is hard for any observer to imagine that its frequency is different at different locations along its path. It would be equivalent to an observer watching a huge train passing through a tunnel in a steady condition and seeing the number of cars that came out in a given time interval differ from the number of cars that went in. This could only happen if cars were accumulating in the tunnel. Even if railroad cars could stretch like rubber and the speed of the train were to be different on different sections of track, the steady state solution is that the number of cars passing a given point per second is a constant. This is equivalent to the statement that the frequency of a light wave doesn't vary from point to point.

The photon theory says the frequency goes up. The wave theory says that the frequency doesn't change. There is a way out of this dilemma. A single observer who looks at the whole wave train does conclude that there is only one frequency. However, observers at different locations examining the frequency of a light wave are using different timing devices. It could be said that the light wave has a higher frequency at the bottom of the building because the clocks there are running slower. This means every conceivable clock, from atomic clocks to biological clocks. If a light wave of a specified frequency is emitted from a source at the top of a building a person receiving this light at the bottom would claim it had a higher frequency. This conclusion is testable by experiment and has been verified. Again it is seen that our present-day notion of time differs from that of Newton.

The above discussion helps us understand certain features of black holes. Suppose a spaceship approaches the event horizon of a black hole. All its clocks are moving slowly, as seen by observers on Earth , although all seems normal on board the spaceship. The closer it gets to the event horizon, the slower the clocks get. The spaceship may be sending signals back toward Earth at what is considered to be a high frequency at the source, but their clocks are running slow. The spaceship may be emitting blue light, but its frequency is reduced to that of red light, as observed on Earth. Everything seems to come to a dead stop as the spaceship reaches the event horizon.

The original conception of black holes made it seem that everything seemed perfectly normal to people aboard the spaceship. The event horizon didn't seem to be observable to passengers on board. Today there is much discussion about this. There is even controversy as to whether space-time actually ends at this horizon. Questions such as these don't have easy answers.

Questions without Answers

Niels Bohr once made the following statement: "If you think you understand quantum theory, and if you are not in awe of it, then you don't understand quantum theory." This may not be an exact quote but it has the essential ingredients. The quantum is shrouded in mystery, and there have been a huge number of questions associated with it. For example, a question that has been often asked is "Is the electron a particle or is it a wave?" It is not hard to think of a beam of electrons as being described by a wave, but what does it mean when a single electron is described by one? Richard Feynman (1918–1988), an extraordinary theoretical physicist , liked to use the double-slit experiment as a way of answering this question. If a beam of electrons of definite momentum were allowed to pass through two closely spaced slits and then strike a fluorescent screen, the pattern on the screen would vary between light and dark regions. This clearly indicates the wave nature of the beam. If the beam intensity were then reduced so that only one electron would get through at a time, only one pixel of the screen would get illuminated at a time. It would look like a random pixel was being illuminated by a particle. On further examination it would be realized that certain regions of the screen never get much illumination while others get a lot. The original pattern from the beam is obtained if the results of the individual pixels are recorded and superimposed. In other words, the hits seem to be random, but the wave nature of the pattern is revealed by the total record. The interpretation is that the wave deals with the <u>probability of</u> finding the electron in a particular region.

In classical physics, the intensity of a beam is obtained by squaring the amplitude of the oscillations. In quantum theory, the probability is also proportional to the square of an amplitude. It is a little more complicated than that because the wave function involves imaginary numbers. You may ask, "How can an electron go through both slits?" The theoretical physicist could reply, "If you try to determine which slit it went through, you would destroy the interference pattern." This is what is observed. However, experiments have shown that if you set up equipment to measure the slit that was traversed by the electron, but destroy the information before it is recorded, the interference pattern can be restored. It is mind-boggling because we like to think in an either/or fashion. We tend to think, "If the electron is a particle, how can it 'know' there are two slits?" If it's a wave, why does it illuminate one pixel? It is as though the electron has the ability to only answer the questions that are asked of it. If you try to determine if it is a particle, it behaves like one.

Einstein respected quantum theory. However, he thought it was incomplete. He had dealt with probabilities before when he dealt with molecular motion in liquids and in gases. The particles had positions and velocities, but they were unknown because of the complexity of the problem. Niels Bohr and Einstein would discuss the theory over and over. Bohr would take the point of view that it would not be possible to gain such information, other than that which was provided by the theory. Einstein would try to find loopholes. Bohr would always be able to come up with a reply. Much as he tried, Bohr couldn't convince Einstein that the theory was as complete as could be. Einstein kept repeating, "I can't believe that

God would play dice with the universe." Finally Bohr got fed up. "Einstein", he said, "please don't tell God what to do."

The probability interpretation is known as the Copenhagen school of thought. There are other mind-boggling interpretations to the theory. One interpretation claims that everything that can happen does happen. It is as though there are an infinite number of histories. Any observer looking backward on the history of things can only see one of these possibilities. Instead of one reality, this "crazy" idea leads to many realities. It is as though there were an infinite number of theaters playing a movie of a possible reality. An individual could only watch the movie in one theater. Worse than that, the number of theaters goes up whenever an event is recorded.

In 1935 Einstein and his two postdoctoral research associates, Boris Podolsky and Nathan Rosen, submitted a famous paper questioning one of the predictions of quantum theory. It is known as the EPR paradox, and it seemed to imply that the theory led to contradictions. The essence of the paper can be grasped by considering a slightly different example than the one used in the EPR paper. Suppose a stationary atom emits two photons of equal energy in two opposite directions. The atom is at rest and has no angular momentum before and after. Photons can be polarized just like a light wave. If one photon is circularly polarized, it carries angular momentum. The other photon has to carry an opposite angular momentum, so it must be circularly polarized also.

Up to this point, there is no paradox. Such correlations exist in ordinary circumstances also. For example, if you and your spouse get two hundred dollars from an ATM, and he hands over eighty dollars to you, then you know that he has one hundred and twenty dollars. Where the difference comes in between the two situations is that the theory predicts that if one of the photons is linearly polarized then the other photons is in a definite state of linear polarization also. The measurements can be made when the photons are thousands of miles apart. The theory says that the two photons are entangled. This is amazing in that most people think that when a photon is emitted, it is in some state of polarization, even if it isn't known what that state is. If a photon is known to be in a definite state of linear polarization, you could be sure of the outcome of an experiment with an instrument that only allows the perpendicular component of polarization to pass. On the other hand, if it is in a state of circular polarization, it is only possible to assign a 50 percent probability to this absorption. EPR asked the question, "How can the decision of what to measure at one location affect instantaneously the outcome of an experiment at some distant location?" It seemed to defy the relativistic requirement that no signal can be transmitted faster than light. On the other hand, does this constitute a signal? The paper seemed to justify Einstein's doubts as to whether quantum theory could be complete. However, many years later, experiments were done that justified the belief in entangled states.

Today the belief in entangled states is so strong that attempts are being made to build quantum computers that rely on quantum states in which many atoms are entangled. If such computers can be constructed, they are supposed to be able to solve, in minutes, certain problems that would take years for ordinary computers. Perhaps Einstein's doubts as to the predictions of quantum theory hastened the development of such computers.

Although quantum theory has had amazing successes, there are still some problems with it. For example, quantum electrodynamics is so successful that certain quantities can be predicted to better than a millionth of a percent. The theory really has an error in it somewhere because certain intermediate steps in a calculation yield infinity. Clever theorists have learned how to get around these difficulties but the theory is not completely consistent. String theory is one of the remedies that has been postulated to avoid some of these difficulties. In string theory particles are not points but strings or membranes that are vibrating in some fashion. Different types of particles correspond to different types of vibrations, resembling different modes of vibration of a violin string. Although the theory is beautiful and impressive mathematically, it seems to predict that there may be many different universes in which the laws of physics may be different in each one. It hasn't produced clear-cut answers.

A more worrisome problem for some theoretical physicists is that quantum theory doesn't seem to be compatible with the general theory of relativity. For most phenomena, the two theories seem to work in different domains. Although there should be no restrictions on the theory, it would be almost impossible to apply quantum theory to the motion of a baseball. Yet it is unusual to think of general relativity being necessary for anything moving at twenty miles an hour. If the Big Bang Theory is correct, there was a time when the universe was sufficiently small to require quantum treatment. It would be nice if the theories could be adjusted so that they became consistent with one another. Much work is currently being devoted to this problem.

The ultimate question seems to be whether physics will ever be a closed subject. The atom has been found to consist of neutrons, protons, and electrons. The first two of these have been found to be made up of objects called quarks. Will quarks be found to be made up of something smaller? There are lots of questions. We don't have lots of answers.

www.ingramcontent.com/pod-product-compliance
Lightning Source LLC
Chambersburg PA
CBHW051445170526
45166CB00001B/124